AF833614

# Agricultural Statistical Data Analysis Using Stata

George E. Boyhan

CRC Press
Taylor & Francis Group
Boca Raton London New York

CRC Press is an imprint of the
Taylor & Francis Group, an **informa** business

CRC Press
Taylor & Francis Group
6000 Broken Sound Parkway NW, Suite 300
Boca Raton, FL 33487-2742

First issued in hardback 2017

© 2013 by Taylor & Francis Group, LLC
CRC Press is an imprint of Taylor & Francis Group, an Informa business

No claim to original U.S. Government works

ISBN 13: 978-1-138-40149-5 (hbk)
ISBN 13: 978-1-4665-8585-0 (pbk)

This book contains information obtained from authentic and highly regarded sources. Reasonable efforts have been made to publish reliable data and information, but the author and publisher cannot assume responsibility for the validity of all materials or the consequences of their use. The authors and publishers have attempted to trace the copyright holders of all material reproduced in this publication and apologize to copyright holders if permission to publish in this form has not been obtained. If any copyright material has not been acknowledged please write and let us know so we may rectify in any future reprint.

Except as permitted under U.S. Copyright Law, no part of this book may be reprinted, reproduced, transmitted, or utilized in any form by any electronic, mechanical, or other means, now known or hereafter invented, including photocopying, microfilming, and recording, or in any information storage or retrieval system, without written permission from the publishers.

For permission to photocopy or use material electronically from this work, please access www.copyright.com (http://www.copyright.com/) or contact the Copyright Clearance Center, Inc. (CCC), 222 Rosewood Drive, Danvers, MA 01923, 978-750-8400. CCC is a not-for-profit organization that provides licenses and registration for a variety of users. For organizations that have been granted a photocopy license by the CCC, a separate system of payment has been arranged.

**Trademark Notice:** Product or corporate names may be trademarks or registered trademarks, and are used only for identification and explanation without intent to infringe.

**Library of Congress Cataloging-in-Publication Data**

Boyhan, George.
Agricultural statistical data analysis using Stata / George Boyhan.
p. cm.
Includes bibliographical references and index.
ISBN 978-1-4665-8585-0 (pbk. : alk. paper)
1. Stata. 2. Agriculture--Statistics--Computer programs. 3. Agriculture--Statistics--Data processing. I. Title.

S494.5.D3B69 2013
635.0285--dc23 2013014245

**Visit the Taylor & Francis Web site at**
**http://www.taylorandfrancis.com**

**and the CRC Press Web site at**
**http://www.crcpress.com**

To Dr. Norton who answered the phone
over the Christmas holidays

# Contents

# Introduction

Stata is a statistical software package that began as a command-line program. A graphical user interface (GUI) was added to the program sometime after its introduction, which has generally been very well executed. It allows beginners and novice users to conduct statistical procedures without having to type commands that can become rather complex with certain models. The command-line approach is never very far away and, as you gain confidence with the program, you will find yourself using it more and more.

The program has matured into a user-friendly environment with a wide variety of statistical functions. A couple of nice features that have dramatically improved usability are being able to have a dataset visible on the desktop, while analyzing data and help menus that indicate where in the menus the specific statistical function can be found.

This book will attempt to introduce the reader to using Stata to solve agricultural statistical problems. Stata, as a general purpose statistical program, has a large suite of commands that are applicable in a variety of disciplines. Based on the number and scope of textbooks available on Stata, it has a strong following in medical, large population, and regression analyses. This is not to detract from its overall capabilities to solve a wide range of problems.

This book provides an overview of using the Stata program. It includes a discussion of the various menus, many of the dialog boxes, and an explanation of how the parts are integrated.

An explanation of how data can be entered into the program or imported is also presented. Surprisingly, for those new to statistical software and analyses, this can be one of the most time-consuming aspects of statistics. Stata has a very in-depth set of capabilities for entering, importing, and manipulating data prior to analyses.

This is followed by a chapter on the simplest of descriptive statistics. An ever-increasing level of complexity as different models and approaches to agricultural statistical problems are introduced follows. One of the biggest changes in Stata is the ability to create graphs. This gives the Stata user another tool in preparing results for presentation and publication.

This book attempts to explain how to use Stata to analyze agricultural experiments. Data that violate the underlying assumptions in many parametric tests must be handled differently. This may involve transformation or the use of nonparametric tests. Various examples from agricultural experiments are covered.

*Agricultural Statistical Data Analysis Using Stata* includes the more important statistical procedures used in agricultural research. Various experimental designs and how to handle them within Stata are discussed. Analysis of variance and covariance applications for agricultural experiments is covered. Post hoc tests and comparisons are covered as well. How to perform regression and correlations with some agricultural examples is included.

The more important nonparametric tests used in agricultural research are also covered—in particular, the use of chi-square for categorical data, such as from inheritance studies.

As mentioned earlier, Stata grew out of a command-line interface, which is still recognizable as part of its foundation. In fact, this command-line interface is one of its strongest attributes because these commands can be organized and executed as a program, which expands the capabilities of Stata and ultimately makes things easier for users willing to devote some time to developing unique programs to solve their particular problems. An introduction to programming Stata is included, which should help users in this area. How to program Stata to extend its usability is also covered. Multiple-range tests are part of Stata, but they will be used as examples on how to implement them in Stata as user-written programs are covered as well. How various programming files relate to one another and how to develop your own programs are also discussed.

Although the programming capabilities of Stata are some of its best attributes, for the occasional user, it may seem quite daunting. This is where the GUI can be a real help. In this book, I present the GUI approach along with the command-line approach, so that the occasional user can use the program without feeling intimidated or thinking they have to climb a steep learning curve.

All of the datasets used in the book are from other texts, from my own research, or made up to highlight a procedure. Where datasets are taken from other texts, the text and page number are listed. These textbooks are listed in the References at the end of the book and all are excellent sources for more information about using the statistics described in this book. In addition, Stata includes all of its reference materials as PDF files with the program. There are links to these files in the online help. These reference manuals have a more in-depth discussion of the specific procedure in question as well as references from the scientific literature.

I try to use the typesetting conventions in Stata's manuals, but won't be presenting commands in as formal a manner. There's no use re-inventing the wheel. For a comprehensive presentation of a particular command, the reference manuals are always there, as is excellent online help both within the program and from the Internet. The figures that present different parts of the program generally alternate between Macintosh® and Microsoft Windows®-based computers. These elements are almost identical between the two systems. So, with that, let's begin.

**George Boyhan**

Data sets available for download at
http://www.crcpress.com/product/isbn/9781466585850

# About the Author

**George Boyhan, PhD**, is a professor of horticulture and an extension vegetable specialist. He has worked for 15 years at the University of Georgia in this capacity and has conducted a wide variety of experiments requiring statistical analyses. Prior to this, he worked at Auburn University as a senior research associate, which entailed designing experiments, collecting data, and analyzing results.

Dr. Boyhan has worked with a wide variety of crops in his career including pumpkins, Vidalia onions, watermelons, cantaloupes, plums, and chestnuts. His current work is with the development of disease-resistant pumpkins, developing watermelon varieties for organic production, and evaluating sustainable production practices.

Dr. Boyhan is an internationally recognized authority on vegetable production. He has given presentations at a number of venues in the United States and internationally. He has published two book chapters, over 40 refereed publications, and many other publications on vegetable production and culture.

"He uses statistics as a drunken man uses lamp-posts... for support rather than illumination."

**Andrew Lang (1844–1912)**

# 1

# General Statistical Packages Comparisons

Stata is a general-purpose statistical program that has some unique features not found in other such general packages. Two other popular general-purpose statistical packages are SAS (Statistical Analysis System) and SPSS (Statistical Package for the Social Sciences). Each of these has its strengths and weaknesses. SAS probably has the greatest user base among agricultural researchers. It is a command-line program that has a GUI (graphical user interface), but it is only available as an add-on. SAS does not maintain the same level of versions across operating systems. So, for example, the latest version available for Windows® is 9.3, while for the Macintosh® it is 6.12, which is not supported in the current Macintosh operating system, and, since I use a Macintosh, well, you get the picture.

SPSS is a statistical package that began life as Statistical Programming for the Social Sciences. Obviously, with such a background, its strong suit is in the social sciences. SPSS, like SAS, does not maintain the same versions across operating systems. The latest available of SPSS uses a GUI exclusively unless you acquire the plug-in for programming.

SAS and SPSS are modular programs with capabilities split over several different modules. This means that certain capabilities may not be available unless you purchase or acquire the necessary module. For a more in-depth examination of all of these general-purpose statistical packages, there are many reviews available online.

Stata takes a much simpler approach to statistical analyses with a single program interface. It, too, like SAS and SPSS, has many parts, but they remain largely unseen by the user. The user does not have to load different modules or pay for additional modules to do specific tasks. Stata does add additional commands, which are available as official updates. There are user-written commands available as well.

Stata also takes the approach of having a tight integration with Internet resources. This is particularly helpful with a high-speed connection. The program will routinely update itself either with your permission or as a background event—your choice. These upgrades are always free within a specific version number. This doesn't sound like much, but the software is routinely upgraded and improved. Searching for help also is integrated with the Internet. Many help files and examples can be accessed from the Help menus. These files may be part of the package of files that were loaded when installed on your computer or they may be on Web sites that the program searches. Stata maintains many of these examples and many are available from third parties.

Stata's commitment to the program goes beyond upgrades. If you need technical help, send your question to Stata and include your serial number; you will get a response within a few days. Not a generic response, but a specific response to your question. They offer a couple of online courses on using and programming the software, which includes many examples in an interactive environment. Their Web site has an extensive bookstore with texts on using both Stata as well as statistical textbooks. They even have a journal, *Stata Journal,* with articles on using Stata to implement various statistical functions.

Finally, unlike other statistical packages that may only offer a limited number of statistical functions, Stata offers a comprehensive set of statistical functions as well as extensibility through its built-in programming language. Stata appears to be committed to releasing versions of their software simultaneously on PC Windows, Macintosh, and Unix® platforms. Stata also takes the approach of having a tight integration with Internet resources. This is particularly helpful with a high-speed connection. The program will routinely update itself either with your permission or as a background event—your choice. These upgrades are always free within a specific version number. This doesn't sound like much, but the software is routinely upgraded and improved. Searching for help also is integrated with the Internet. Many Help files and examples can be accessed from the Help menus. These files may be part of the package of files that were loaded when installed on your computer or they may be on Web sites that the program searches. Stata maintains many of these examples and many are available from third parties.

### Program

Stata is available on the three major operating systems: Windows, Macintosh, and Unix. In addition, there are several flavors of Stata available. These include Stata/MP, Stata/SE, Stata/IC, and Small Stata. These versions differ in the type of machine they can run on and the size of datasets they can handle. Stata/MP is for multiprocessor machines, while Stata/SE is for single processor machines. Both of these are considered the professional versions of the software and both handle the largest datasets.

Stata/IC, which was formerly known as Intercooled Stata, is the intermediate-sized program, while Small Stata handles the smallest of datasets and is the slowest of the versions. Small Stata is primarily used for educational purposes. If you haven't already purchased a Stata program, you should know they are priced differently with the greater capacity programs obviously costing more. In addition, if you haven't purchased the program, check with your institution. It may have a site license agreement with Stata that would make the program available to you at a greatly reduced price. Finally, pricing is different based on the type of purchaser.

Printed documentation also is available. This documentation includes manuals on using Stata with specific operating systems: a Base Reference Manual (four volumes) or reference manuals on specific subjects, such as a graphics manual, data management manual, programming manual, survey data manual, as well as several others. This documentation comes with the program as PDF files and is linked to the Help menu.

Obviously, such an extensive set of manuals is not meant to be read through, but is to be used as a reference source. Although I will be going through many of the basic functions of the program to start with, it's a good idea to read through the *Getting Started with Stata** manual for your specific operating system. This manual is available for either Windows, Macintosh, or Unix depending on which version

* Stata Press. 2011. *Getting Started with Stata.* College Station: Texas.

of the software you buy. It is a great introduction to the program that will help you get a feel for how it works and gives you an opportunity to work though some examples.

*Windows and Menus*

There are several windows in Stata, each with a unique and useful function. All of these windows are accessible under the Windows menu. This brings up an interesting point about using Stata. With the number of windows and available information, having a large monitor can be very helpful. With a large monitor, you can view several windows simultaneously, which makes it much easier to use. The Command, Results, Variables, and Review windows are integrated into a single window, referred to here as the Main window. These areas (i.e., Command, Results, Variables, and Review) are often referred to as *windows* and are listed separately under the Window menu.

In previous versions, the Results window appeared with a black background in the default setting. This is now referred to as the Classic setting in the Preferences menu. The Classic view is particularly nice because different colors are used on a black background for the various types of output. This can be particularly helpful when learning the program. This window is where all of the results of your analyses will appear as well as echoing commands you type in or initiate from the GUI dialog windows. This window has a reasonably large buffer so you can scroll back to look at previous analyses and commands. This buffer is not unlimited, however, so eventually results will no longer be visible as more and more information is added.

Figure 1.1 shows the Main window right after you have opened the Stata application. There are several pieces of information displayed in this window upon startup: the version number, company contact information, and the license information. The blue texts are live links, which can be clicked to go to Stata's Web site or to send an email to Stata, which requires an Internet connection.

Text will appear differently in the Results window depending on its source. The default output is black, black/bold, red, and blue with each representing something different. Text in black/bold represents the command and this information will change depending on the command and the dataset in memory. Black text is for labels to indicate

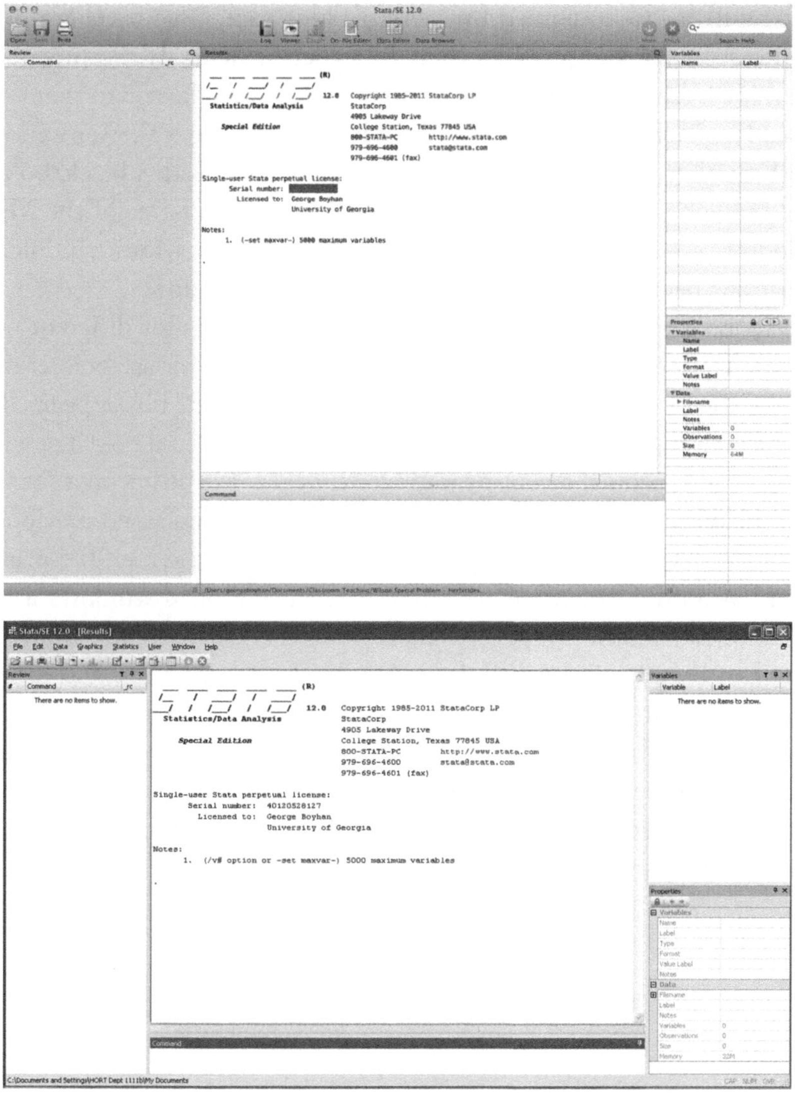


**Figure 1.1** The Main window immediately after opening as it appears on Macintosh (top) and Windows (bottom) computers.

what results (black/bold text) are. So, for example, analysis of variance labels for sum of squares, degrees of freedom, etc. will appear as black text. Black text changes based on the command, but will always label the same things within a command. Red text indicates an error—a command was entered incorrectly or used inappropriately depending on the situation or variables selected. Usually an error message (red text) will be accompanied by a link in blue text. Blue texts are links and can be clicked just like in an Internet browser. If the link (blue text) is a Web page, it will open your browser and take you to that location. In general, however, these blue links will open a Viewer window with further explanations concerning the error. Finally, black/bold is used to echo what has been typed in the Command area of the Main window, which appears as the lower portion of the Main window, or what has been entered into a command dialog window.

At the top of the Main window are several icons for different purposes. To find out what these icons are for, roll your mouse pointer over one of the icons for a few seconds and a yellow "about" box appears. The first icon is for opening data files. If you press the icon and hold it down, a drop-down menu of recently saved files appears. The next icon is for saving the dataset in memory. If the dataset has not been saved previously, a standard save dialog box appears for you to save the file. The printer icon has a drop-down menu with all the current open windows listed. Selecting a window brings up a small dialog box with several parameters that can be set prior to printing, including a header, user, and project fields (Macintosh only). Other parameters include Stata fonts and colors, which are available from a drop-down menu (Macintosh only). You can select to print either the Results window or any open Viewer windows. These are selected by holding down the Printer icon until a drop-down window appears with window selections (Figure 1.2).

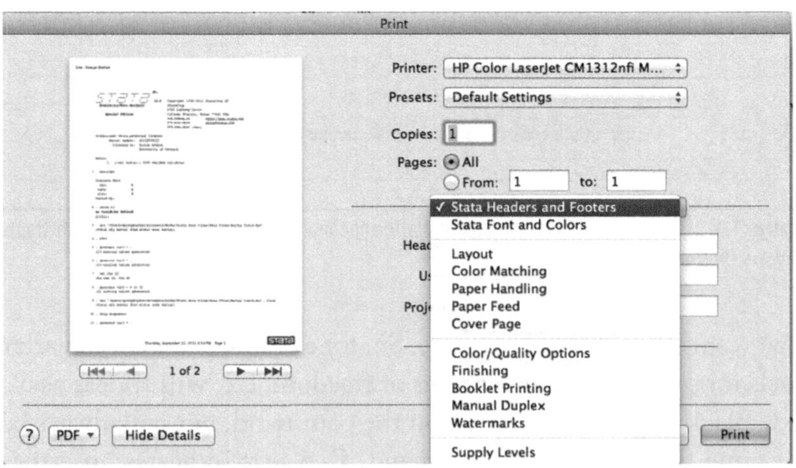


**Figure 1.2** Printer dialog box with drop-down menu showing Stata selections on a Macintosh computer.

The next icon is the Log icon (it's suppose to look like a little log book). This is where you can turn on a log (Begin) so that everything you type, as well as the results, is entered into a file. You also can Suspend and Resume your log and finally close the log file. You can view your log or any log for that matter by selecting the View ... option under the Log icon. On a Windows computer, selecting the Log icon the first time opens a dialog box for saving the log. Subsequent selections of the Log icon will bring up a dialog with selections for viewing a snapshot of the log file, closing the log file, or suspending the log. These log files will appear in a Viewer window when you open them. Log files can be saved as either .smcl or .log files. The former is Stata's markup and control language and the latter is a text file that can be opened by any word processor or text editor.

The eye icon is for opening Viewer windows. You can open a new Viewer window or, by holding down the icon, select any Viewer window that is open. Finally you can close all of the open Viewer windows at once.

The next icon looks like a little graph and will bring the Graph window to the front, if a graph has been constructed; otherwise it won't work. If there are one or more graph windows open, this icon will allow you to select a Graph window or Close All Graphs.

The next icon that looks like a page with a pencil is to start a Do-File Editor Window. Stata is a fully programmable statistical package and the Do-File Editor is where this is accomplished. You can enter lists of commands in the Do-File Editor and Stata will execute them in sequence. Further, these files can be saved, so you have a sequence of commands that you can use more than once. The programming capabilities of Stata go far beyond just a simple sequence of commands and that will be covered in greater detail in Chapter 7. Suffice it to say that just having the capability to execute a sequence of saved commands can save a lot of time and be a powerful tool in analysis. If you have more than one Do-File window open, clicking and holding the Do-File Editor icon will show a list of currently open Do-File windows, which you can choose to bring to the front. Each Do-File is a separate tab in the Do-File Editor window. The Data Editor can be opened by clicking its icon.

The next icon is the Data Browser, which opens the Data Editor window, but no changes can be made to the data in this view. This is to help prevent you from inadvertently changing data in the Data Editor.

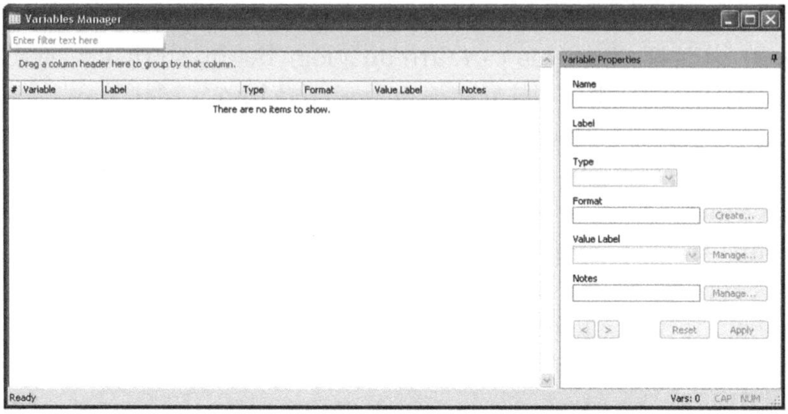


**Figure 1.3** Variables Manager window as it appears on a Windows computer.

On a Windows computer, the next icon is the Variables Manager. This opens a window listing the variables in the dataset and has entries for changing variable names, controlling the format, changing the data type, and adding labels (Figure 1.3)

The More icon clears the -more- condition, much like hitting the space bar would. Finally, the red X icon on a Macintosh or a blue X on a Windows PC is a break button to stop a command, program, or Do-File before it has completed executing. This is handy if you encounter an error or just wish to stop the current program action. That is an overview of the various windows and how they function.

The Variables and Properties region of the Main window have several additional features. The down arrow in the Variables header region can close and open the Properties region below on a Macintosh. On a Windows PC there is a push pin icon that does essentially the same thing. In addition, the magnifying glass icon (Macintosh) or the funnel icon can be used to find or list specific variables. In the Properties region is a small lock icon that can be on (locked position) or off (unlocked position). When it is locked, no changes can be made to the variables. There is also a forward and backward arrow to cycle through the listed variables.

The Properties region is used to add labels to variables, set up value labels, and change numerical types (i.e., float, double, long, integer, or byte). The filename is listed here, as well as the file label and any notes. Additional information about the size of the dataset also is listed in this region.

All of the regions of the Main window can be resized for convenient viewing. In addition, under the View menu on a Macintosh is the Layout submenu with selections for rearranging the Main window as to placement of the Command, Results, Variables, and Properties regions. This same functionality is available on a Windows PC by simply dragging the window region to a new location.

Viewer windows are where information about commands or statistical procedures appear. There is an extensive online help system built into Stata. In addition, if you have an Internet connection you can simultaneously search Web resources for additional help. There can be more than one Viewer window open at a time, so multiple pieces of information can be available simultaneously. You can open a new Viewer window from under the Window menu. The blue texts within a Viewer window are links to other information. This information may be on your computer or, if you have an Internet connection, it can be retrieved from remote sites.

At the top of the Viewer window are several icons, buttons, and an input field (Figure 1.4). The input field is where you would type "help" with a Stata command or "search" with a term you are looking

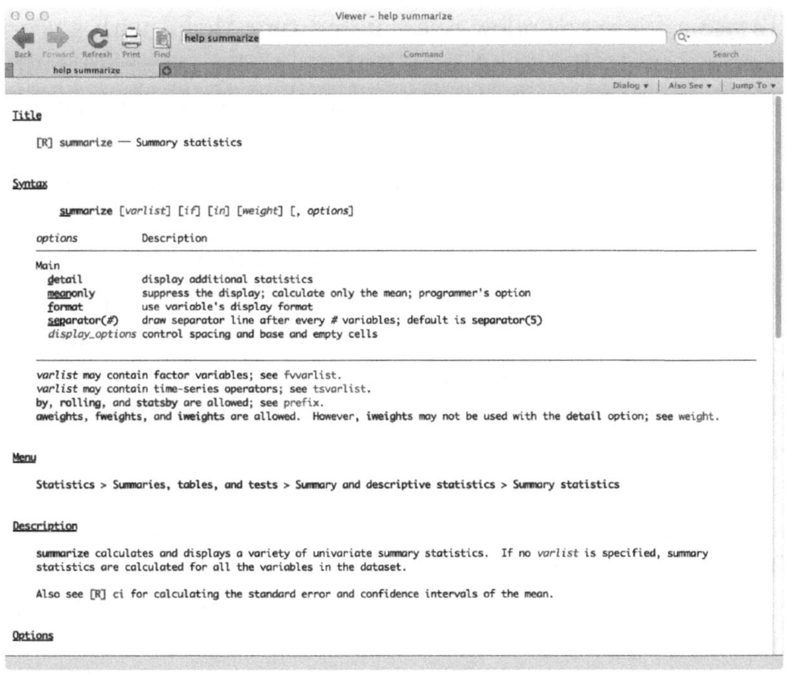


**Figure 1.4** Viewer window on a Macintosh.

for that is not a Stata command. In addition, there are left and right arrows. These are used to move backward and forward through Viewer screens. So, for example, you may have looked for help on several different commands and these arrows allow you to quickly move back and forth between screens. It works exactly like equivalent buttons in your Web browser. The arrows in a circle are to refresh the current screen, again just like in a Web browser. The icon of a printer, as you would expect, is to print the window contents.

The Find icon can be used to search for text in the current window. When this icon is selected, a search field is available at the bottom of the window. Type text you are looking for within the current window and all entries within the window will turn yellow. You can move between each entry from your keyboard.

In addition, the Viewer window has three additional buttons labeled Dialog, Also See, and Jump To. The Dialog button takes you to the dialog box used for the currently listed command. The Also See lists where more information can be found in the documentation either built into the program or the PDF files that came with the program. The Jump To jumps to specific topics in the current window.

To use a Viewer window select it and type "help" with a specific Stata command. The window will then display information about using that specific command. Along with the Help command, you can type in "search" followed by a term that is not a Stata command to see what information is available about that term. There is an additional "search" function in the upper right hand of the window that can be used for searching documentation and frequently asked questions, searching net sources, or searching both. For example, searching "transformation" will list a variety of Stata commands associated with this term. In addition, a variety of questions about this term with associated Web pages also are displayed. Finally, additional commands that may not be installed on your computer are listed with links to their location for downloading. These downloadable commands usually come with a downloadable help file as well.

The Viewer window also can have several tabbed items available at the same time, much like an Internet browser. Additional tabs can be added by the user.

Viewer windows are where log files are displayed as well. Within Stata, you can turn on a log that saves everything you type as well as the results to a file. If you wish to view one of these logs, it will appear in a Viewer window when loaded. I will have more to say about log files later.

The command entry region at the bottom of the Main window is where all of the commands are typed for manipulating data and making statistical calculations. You type a command here and when you hit return, and assuming there is no error in what you have typed, both the command and the results appear in the results region above.

The next area of the Main window is the Review region. This is where all the typed commands appear as well as error codes if the command is incorrect in some fashion. The Review has an error column that has the heading _rc, for return codes. You can adjust the width of this region by sliding the vertical bar between this region and the Results region. The width of the _rc column also can be adjusted in the header. Finally, the Review region has its own search function. Click on the magnifying glass icon at the top of this region. An interesting feature of this region is, when clicking on a previously typed command, it will then enter it in the Command region. Then you just have to hit return and the command is executed. Although I've been talking about typing commands to get results, you can use the menus to select your command. A dialog box appears and you fill in the parameters and hit OK. The command is entered in the Review area just as if you typed it in the Command region.

The next region of the Main window is the Variables list where all of the variables in the currently loaded dataset are listed. In addition, any labels associated with a particular variable are listed. The variable type and format are below the list in the Properties region of the main menu. Selecting the column to the left of a variable in the Variables list will automatically enter it in the Command region. This can be helpful if you are executing a previously entered command, but are changing one or more of the variables.

The Data Editor is a spreadsheet-like window where data can be entered (Figure 1.5). The Data Editor can be opened for editing or browsing by selecting one of the two icons in the main window (see Figure 1.1). For example, census data or a database of important medical information, whose integrity should not be compromised, can be opened for browsing and not be inadvertently changed. This is rarely the case in agricultural statistics where planned experiments of comparatively smaller datasets are involved. In addition, the Data Editor

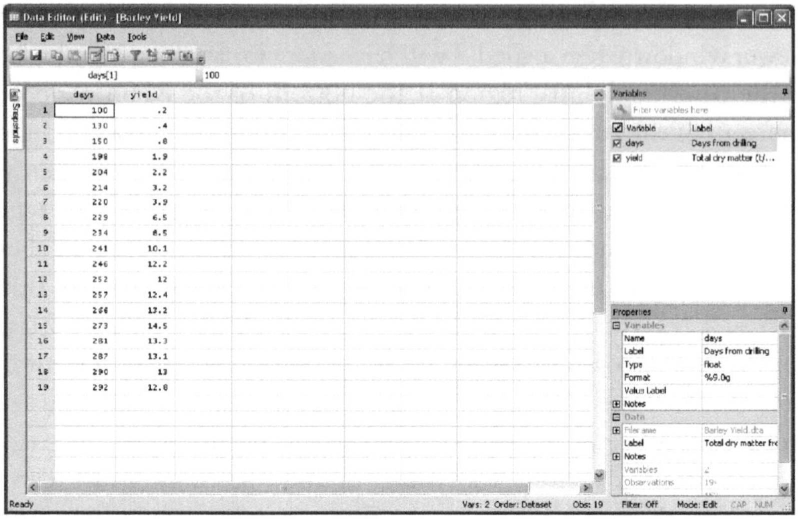

**Figure 1.5** Data Editor window as it appears on a Windows PC. It will appear somewhat differently on other operating systems.

can be invoked by typing **edit** in the Command area of the Main window. The Data Editor also can be opened so that changes cannot be made by typing **browse** in the Command window.

The Data Editor works just like any spreadsheet. If you are familiar with Excel, the Data Editor works in a similar fashion where data are entered in cells defined by the row number and column heading. In Stata, as in most statistical software, the rows are referred to as *cases* or *observations*, while the columns are referred to as *variables*. The selected cell will appear with a black rectangle. The Data Editor is not capable of producing a noncontiguous dataset; therefore, if you select a cell by itself and enter a value, the Data Editor will enter missing values in all the empty cells from the first cell (row 1, column 1) to the cell in which you have entered data. The missing data will appear as periods (.).

At the top of the Data Editor are several buttons. One such button is the Filter button. Data can be filtered so that specific cases or variables don't appear. This does not affect analysis, however, but doing an analysis on a subset of the data is not a problem as most commands allow this.

The Variables button is used to hide or show the Variables and Properties region on the right of the Data Editor window. The Properties button hides or shows the Properties region of the window.

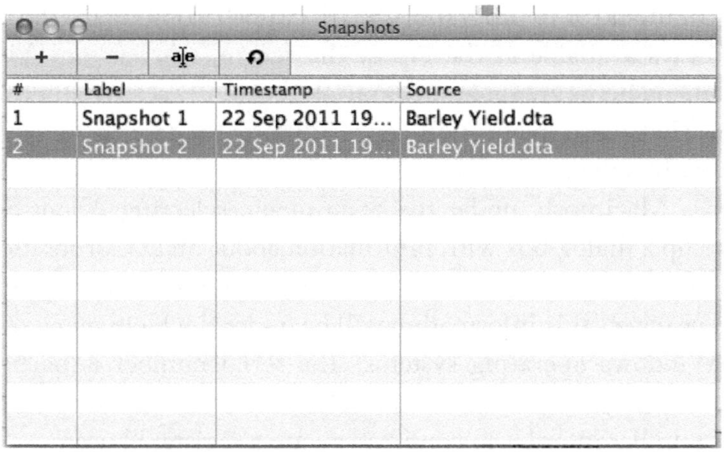


**Figure 1.6** Snapshots window on a Macintosh.

The Snapshots button brings up a dialog box that allows you to take a "snapshot" of the current dataset (Figure 1.6). On a Windows PC this will slide out from the side of the Data Editor and not be a separate dialog box. This can be helpful if you are interactively changing the dataset; for example, using the **`collapse`** command to look at or analyze a portion of the data. From the Command area entering **`preserve`** and **`restore`** works in a similar fashion. The + and − icons work as would be expected for adding or deleting snapshots. The icon next to these is for changing the snapshot's name and the last icon is for restoring the dataset.

*What's on the Menu?**

Let's take a moment and look at the different menus and what functions are available from them. As I mentioned previously, Stata is a general-purpose statistical package with many capabilities that may not all be applicable for agricultural research, so I will not be giving a detailed accounting of every menu item. Instead a quick overview of general capabilities is in order. Stata uses many menu items much like other programs from within a GUI. In some cases, however, Stata invokes menus in a nontraditional way, which comes from its heritage

* Items described here may appear under different menus on a Windows or Unix computer.

as a command line program. On Macintosh computers, the menus are always available at the top of the screen, whereas on Windows PCs, menu items are integrated into the currently active window. This means that these menus will appear differently depending on which window is active.

On a Macintosh, under the Stata menu, selecting About Stata… brings up a dialog box with information about Stata Corporation and how to contact them, the version of Stata you are running, and the serial number. This information will be under the Help menu on Unix and Windows operating systems. The serial number is particularly important if you need technical help from Stata. They require your serial number in order to confirm you are a registered user.

The Preferences menu is located under the Stata menu on a Macintosh, and under the Edit menu on a Unix or Windows PC. There are several selections you can make. The first is General Preferences, which brings up a window with several items you can select or change to determine how Stata will react (Figure 1.7); for example, how data are saved, how searches are handled, which directory to use, etc.

The Graph Preferences brings up a dialog of items that affect the color, font, printing, and clipboard when dealing with graphs. On a Macintosh, there is only one preferences dialog, which opens to the General Preferences or Graph Preferences based on the menu selection, but once the dialog is open you can switch back and forth from the General to Graph Preferences by checking the icons at the top of the window.

There are other icons at the top of the Preference window on a Macintosh for changing other aspects of Stata. The Windows Preference dialog uses tabs. The Do-File Editor icon is used to make changes to how Do-File windows and programs behave. The Syntax Highlighting icon is to set colors for various programming elements in the Do-File Editor. The Windows icon allows you to set parameters for the various windows available in Stata. Finally there is an Internet icon that can be used to set up a proxy server with user name and password as well as determine how often you wish Stata to be updated. As mentioned previously, Stata is tightly integrated with Internet connectivity. Stata Corporation offers frequent updates to its software that can be downloaded and installed automatically. This is a great feature and I encourage you to take advantage of it.

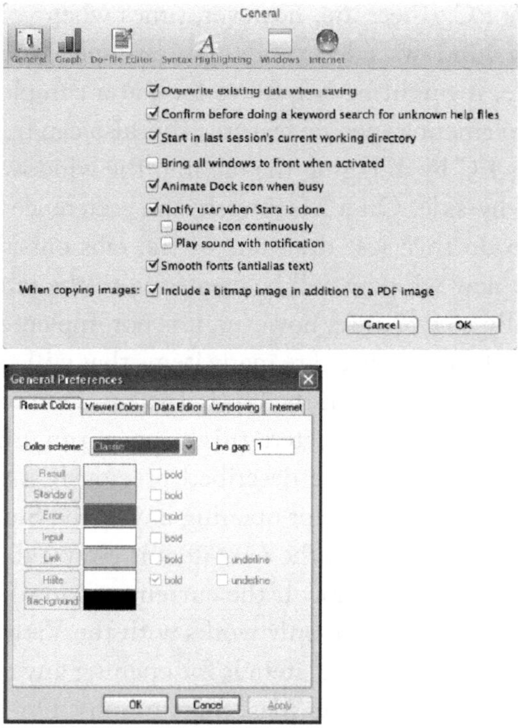


**Figure 1.7** General preferences window on a Macintosh and Windows PC.

Also under Preferences is the Manage Preferences submenu on a Macintosh, which has Manage Preferences…, Save Preferences…, Factory Settings, and Factory Window Settings. (These items may appear slightly differently or not at all under Unix or Windows computers.) You can set up Stata's windows, fonts, colors, etc., and save this as a custom preference file. These files are saved with an .rc extension in the Stata preference folder and can be opened at any time. If you wish, you can reset the Stata program to both the factory settings and the factory window settings on a Macintosh. On a Windows PC, there are several predefined window settings under the Load Preferences Set submenu. This includes the Widescreen Layout (default), Combined Layout, Compact Layout, and three Presentation layouts. Finally, on Windows PCs, there is a Reset File Associations submenu.

Opening new windows in the Do-File Editor or Viewer can be set from the Preferences to open them as new tabs or new windows on a Macintosh. Opening new windows as tabs can help keep your screen from getting cluttered with too many open windows. This is the default on a Windows PC. There are, however, times when you may wish to view two such windows side-by-side. For example, when working on a new Do-File, it might be helpful to look at a complete Do-File to see how to implement a specific feature. This also can be accomplished on a Windows PC by dragging the tab into the window to show both Do-Files side-by-side. On a Macintosh, the preferences don't have to be changed to do this; just drag one of the tabs outside the current window and a new window will be created with the tabbed item. Try it; this is a really nice feature; however, it is not implemented in Unix.

Under the File menu there are many items that will appear familiar to you if you are familiar with the GUI. The first item on a Macintosh is for a New Do-File. As expected, this brings up an untitled Do-File Editor window, which I have described previously. On a Windows PC, the first item is Open… for opening any of the Stata file types.

On Macintosh computers, the next item is New Tab, which adds a new tab to the current window if the current window is a Do-File or Viewer window. This feature only works with the Viewer window on a Unix computer. The Open… item is for opening any of the different Stata files, which include data files, Stata graphs, Do-Files, etc. The Open Recent menu item has a submenu of recently opened datasets. This assumes there are any recently opened datasets. If you are using the program for the first time or have reset the preferences, no submenu will appear.

Other items not on Windows PCs include Open Recent Do-Files. As would be expected, recent Do-Files are listed in the submenu. This is not implemented on Unix computers. Do-Files will have a .do extension. Other files that may appear under this menu include .smcl and .dct files. The .smcl files are output files from Stata in Stata Markup and Control Language. It is not advised to open these files in a Do-File Window because all of the control codes appear rather than the expected formatted output. The Insert File… menu item will appear dimmed unless a Do-File Editor window is open, in which case you can use this to insert a file into the Do-File Editor.

The Close item does just that, closing the current window, and the Close Tab closes the current tab in windows that support tabs. The Close Tab item is not available on Unix or Windows computers. The next two menu items, Save and Save As…, are for saving dataset files if any of the windows are active except the Do-File Editor window, in which case, these menu items will allow you to save the Do-File (extension .do). Datasets are saved with the .dta extension. They work just as they would in any other program within the GUI.

The View… menu item is implemented a little differently than you would expect for a function that opens files. When invoked, a dialog appears that asks for a file or URL path. You can select the Browse… button and a normal file dialog appears, which works as you would expect. View is for viewing do, ado, and smcl files, to name a few. A URL can be entered to access a specific Internet page. If you type a URL address, it will open the Web page as html in a Viewer window. If you have a URL for a particular Stata program, you can view it directly in a Viewer window, which can be helpful.

The Do… menu item is for opening previously saved Do-Files. Once open it can be run, which makes the file available for execution. Executing a Do-File is done by typing **do** followed by the filename in the Command region of the Main window along with any parameters the file requires.

The Filename… item from the File menu is used to select a filename that is going to be part of a Stata command. Some Stata commands require a filename and this menu item quickly allows you to find and select the needed file. Filename will insert the correct pathname with the necessary quotes into the command when selected.

The Change Working Directory… menu item allows you to change the working directory. The working directory is where Stata looks for files you have saved. By changing the working directory, it makes it simpler when typing a command that requires a filename. Ordinarily you would have to type the entire path name to the file, which can become tedious. With the working directory changed all that is needed is the file name. The working directory is where Stata will also look for ado-Files that you (or others) have created. Ado-Files are do-Files that automatically load and run when invoked. Along with those .do and .ado files stored in the working directory, many other of these ado-Files are part of Stata's official updates and are stored in specific folders that Stata knows about and can find when a specific command written as an ado-File is invoked. This is a good reason not to mess with the Stata files that have been installed on your computer or the hierarchy of their folders. These types of files will be discussed at length in Chapter 7 (Programming Stata).

If you have created a graph, the next File menu item, Start Graph Editor, will invoke the graph editor, which allows you to make changes and customize the graph on a Macintosh computer. This menu item is not available on Unix or Windows PCs. I will have more to say about graphing in a later chapter.

The Log item in the File menu is for starting logs, which record all of your inputs as well as the results of commands. In other words, a complete record of your session can be recorded and saved. There are two types of files that can be created. One has a .log extension and is a simple text file that can be opened by any program capable of reading a text file, such as a word processor or text editor. The other type has a .smcl extension that is in Stata's own format and is best viewed from within Stata. All of the error codes maintain their red color, and the links (blue color) are still active in these files when viewed in Stata. In addition, all the formatting remains the same.

The Log menu item has a Begin submenu, which is how a log is started. When started, you have the choice of creating either a .log or .smcl file. The ability to create one or the other file type is not available on a Unix computer. You also can suspend logging with the Suspend submenu and, of course, resume with the Resume submenu. You may wish to do this when you get off on a tangent, but I digress.

When you are finished with logging your session, you can select the Close submenu, which will close the log file. This file then can be viewed within Stata or, if it's been saved as a .log file, with any program capable of opening a text file.

Finally, the Log menu has a Translate submenu, which allows you to translate .smcl files to .log files and vice versa. This can be helpful in getting results into other programs for publication, etc.

The next command under the File menu, Import, deals, as you would expect, with importing data into Stata. The first command is for importing Microsoft Excel® files (.xls, .xlsx). It allows you to examine an Excel workbook, select specific worksheets, as well as cell ranges, and import the data into Stata. The next four items are to import text files in various formats. The first of which imports text files created in a spreadsheet program. Importing text in a fixed format is for files that have fixed column spacing for each variable, but no specific delimiter, such as a tab or comma character. The next item, "Importing text in fixed format with a dictionary," is a unique method of importing. It consists of two files, the text file with the data and a separate dictionary file, with a .dct extension that describes the data for the purposes of importation. Finally, for text file importation, there is an item for importing an unformatted text file.

Importing SAS XPORT, ODBC data source, and XML data also are for importing data into Stata, but deal with importing from another statistical or software package, SAS XPORT from SAS, from a database source (ODBC—open database connectivity), or from any application that supports the open source XML format.

The Export menu also has selections for exporting Microsoft Excel files (.xls, .xlsx). There is a Comma- or tab-separated data, Text data (fixed- or free-format), SAS XPORT, ODBC data source, and XML data, for exporting data files.

As mentioned previously, Stata maintains tight integration with the Internet. This is evident with the next menu item under File, Example Datasets..., which when selected brings up a Viewer window with links to Stata example datasets. One link is to datasets that were loaded on your computer when Stata was installed. As you read through Stata's documentation, it refers to these example datasets to illustrate Stata's capabilities. Clicking on the link Example datasets installed with Stata will bring up a list of datasets used as examples. You can then select one of these datasets to load or click on the "describe" link to see a description of the dataset, which will appear in the Results area of the Main window. On Windows PCs, after the Example Datasets... is the Recent Datasets menu item, which does not appear on a Macintosh.

The Page Setup... item is just that, a command to set page printing criteria, such as paper size, printer selection, orientation, and scale. It is not available on Unix or Windows PCs.

Finally, under the File menu is the Print option. On Windows computers, the Print item appears after the Export menu item. Stata can print out the contents of the Results area of the Main window, any Viewer window, and any Do-File Editor window. A submenu under the Print menu lists the currently available windows for printing. Again expect to see slight differences based on the operating system you are using.

Selecting Print for Results or Viewer windows brings up an Output Settings dialog on a Macintosh, where several parameters can be set

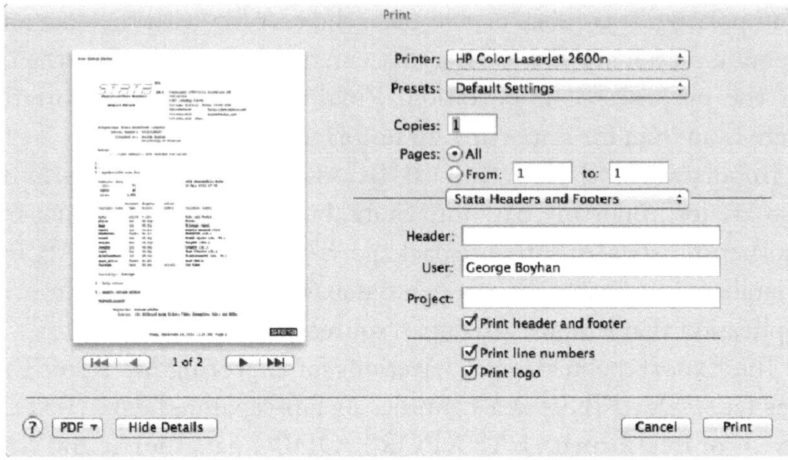


**Figure 1.8** The printer dialog box with several parameters that can be set in Stata on a Macintosh.

for printing. This includes printing line numbers, a header, and printing the Stata logo. In addition, you can include a unique header, name, and project (Figure 1.8).

On Windows PCs under the Edit menu are the menu items Copy, Copy Table, Copy Table as HTML, Copy as Picture, Paste, Table Copy Options…, Find, Find Next, and Preferences. On a Macintosh under the Edit menu are commands for Cut, Copy, and Paste, as well as Undo and Redo. Undo and Redo are not available with Windows and Unix computers. Data or text can be copied from any window in one of several different formats. For example, the Copy command just copies as text and it is pasted into another program exactly as is. If the text is copied with Copy Table (and it is in a table format) when it is pasted into another program, it will have tabs between the columns rather than spaces. This is particularly useful when moving information into, say, a word processor or spreadsheet program for final presentation. This makes formatting the final table much easier. You also can copy the information as an HTML table with the Copy Table as HTML command. This is useful if the information is going to be presented on a Web page. In order to use the Copy as Picture menu item, you have to select Include a bitmap image in addition to a PDF image in the General Preferences on a Macintosh. This allows selected items to be moved to other programs as bitmapped files. This is not available on a Unix computer.

The Paste command operates as you would expect with information copied from other programs pasted into Stata. Data can be pasted into the Data Editor window that includes the column titles, if present, and Stata, which will enter the data into the cells. Stata asks if column titles are present and places that information in the gray column titles row at the top if needed.

In addition to the Paste command is the Paste Special…, which is available for pasting into the Data Editor. This menu item gives you more control over pasted material including what is used as delimiters between data and how sequential delimiters and double quotes are handled.

The Clear item under the Edit menu is used to clear selected commands from the Review window on a Macintosh. Select a line or several lines in the Review region of the Main window and then select clear.

The Table Copy Options… is used to remove vertical lines from a table. Say you have created a table in Stata and want to copy it to another program. In Stata, there may be vertical lines present that might be difficult to remove once moved to the new program. With the Remove All or Smart Remove selected, these vertical lines automatically will be deleted upon pasting into the other program. This is not available on Unix computers.

On a Macintosh, Select All selects all the text in the Results or Do-File Editor window, which can then be copied. The Select Line and Delete Line do just that in the Do-File Editor. These items will appear dim if they are not useable in the current window.

The Find item, under the Edit menu, has several submenus on a Macintosh, which are used within the Do-File Editor; otherwise they appear dimmed. These items are available on Windows PCs from within the Do-File Editor. When Find is selected with a Viewer window as the frontmost window, a Find toolbar appears at the bottom of the Viewer window. This also is available within the Viewer window on Windows PCs. With this toolbar active, a keyword search can be initiated to find the word searching forward or backward in the current document. This should not be confused with the Command and Search fields at the top of the Viewer window that can search from Stata help files on your computer or over the Internet.

In the Do-File Editor, selecting the Find icon brings up a dialog box with several options. You can find, find and replace, and have the option of replacing items one at a time or all at once. There are checkboxes for ignore case and wrap around. The Wrap around checkbox allows the search to continue at the beginning once the end of the document is reached. Another editing feature of the Do-File Editor window is the ability to select text and then drag it to another location in the window. This is a handy feature for editing Do files. This feature also can be used to copy text from one location to another by holding down the option key on a Macintosh or control key on a Windows PC as you drag the text. This makes a copy rather than just moving the text.

Under the Find submenu, there are several submenus with keyboard shortcuts that can make finding and replacing text within a Do-File Editor window quick and easy. There is a Find Next, Find Previous, and Find Selection. In addition, there are submenu items for bookmarks that can be used in the Do-File Editor. Bookmarks can be set for lines of code and can be quickly found again. These submenu items are Next Bookmark, Previous Bookmark, and Toggle Bookmark.

Line numbers in the Do-File Editor can be found with the Go to Line… submenu. In small Do-Files, this may not be important, but in larger files it may be, particularly if you are looking for an error in the code.

The last two submenu items under Find menu are the Balance Braces and Match Braces. Balance is used with [], {}, () brackets and selects all the text in a Do-File Editor between any pair of these. In programming, this can be an important tool to see what a particular subroutine encompasses. The Match Braces submenu has a similar function only it just moves the cursor to the matched bracket. To use this command, the cursor must be in front of a specific bracket. The usefulness of these commands will become more evident as you do more programming.

The next menu item under Edit is the Advanced menu item. This is used with the Do-File Editor to indent or unindent lines, make selections upper or lower case, show or hide nonprinting characters, and choose whether to wrap lines. The last two items are not available on Windows or Unix PCs; however, on Windows computers, there are View Whitespace and View End of Lines, which are functionally the same.

The final two commands under the Edit menu are used when a graph window is open. The first allows you to rename a graph. The last command under the Edit window is Apply New Scheme, which is used to set a new color scheme. These are available from within a Graph window. There are several predefined color schemes to choose from including one for *The Economist* and another for *Stata Journal*. Also, you can look for other schemes by typing **`findit scheme`** in the Command window, which will search the Internet for additional schemes. The last command on Unix and Windows computers will be the Preferences item.

The next menu is the View menu, which is only available on the Macintosh computer. Under this menu are several menu items for dealing with the various windows available in Stata. The Data Editor item has features for entering the Data Editor to edit or browse, manage value labels, filter the data, select the Variables or Properties regions of the Main window, and manage snapshots.

The Do-File Editor lets you execute the program in the current Do-File Editor, execute the program from the cursor location, or run the program. I will have more to say about this in Chapter 7 on programming.

The Graph Editor item has features available when the Graph Editor window is the current window and the graph is in editing mode. There are submenu items for graph objects, the entire graph, and to use the recorder function. Various tools can be selected including the Select Tool, Adding Text Tool, Add Line Tool, Add Marker Tool, and Grid Edit Tool. Finally, the Object Browser can be shown or hidden.

I will skip the SEM Builder because I won't be covering it in the book. The Viewer menu has items to move backward and forward through viewer screens as well as for refreshing the screen.

The Layout menu item is used to rearrange the regions of the Main window. The default view is the Widescreen View, which can be changed to the Combined View where the review, variables, and property regions are on one side of the window. In addition, the Command and Results regions can be swapped, as can the Review and Variables regions.

The View menu also has selections for making the text bigger or smaller in the currently open window. You can hide or show the toolbar at the top of the current window. The toolbar can also be customized by selecting Customize Toolbar… . The toolbar customization is unique for each type of window. The last two items under the View menu are the More and Break menus that are only available when a command or program is running. The More menu item can be selected when the currently running command pauses before bringing up the next set of results to continue to the next screen. The Break menu will stop any currently running program. So, if you have written your own program and there is a problem, selecting Break will stop the program. In addition, the Break menu can stop additional results from scrolling in the Results window. All of these menu items under the View menu are available on the Macintosh computer only. Most of this functionality is available in other places in the Windows or Unix versions of Stata.

The next three menus—Data, Graphics, and Statistics—are the heart of Stata's real purpose and functionality. Because this book covers just agricultural statistical procedures, not all of the commands available under these menus will be used. Commands appropriate to agricultural statistics will be discussed as appropriate for the topic in upcoming chapters.

The User menu is used for commands users develop for their specific purposes. This menu does not have to be used for user-developed commands, but may be convenient for often-used commands or commands that are to be shared with others. The added menu items would, in practice, invoke a custom designed dialog box in which the user would add the necessary input(s), which would then execute the user-created command. Hence, like a built-in command in Stata, you can write programs with a selectable menu item and custom dialog box along with a command for end users to utilize.

The next menu is the Window menu where all of the Stata windows can be selected in turn. This includes the Command, Results, Review, Variables, Properties, Graph, Viewer, Data Editor, Do-File Editor, and Variables Manager. These are the only menu items under the Windows menu on Windows PCs. Currently available windows are listed at the bottom of the Window menu on a Macintosh computer. In addition, on Macintosh computers, the Window menu has items for enlarging the current window to fill the screen (Zoom) as well as minimizing windows (Minimize). The Bring All to Front menu does just that and brings all the open Stata windows to the front of your screen. The Select Next View and Select Previous View will change the active region of windows that have such regions (i.e.,

Main window and Data Editor). These menu items change to Select Next Tab and Select Previous Tab when the graph window is the frontmost window.

The final menu item is the Help menu. This menu includes

- Search field (Macintosh only)
- PDF Documentation
- Advice
- Contents
- Search...
- Stata Command...
- Graph Editor (Macintosh only)
- What's New
- News
- Check for Updates
- SJ and User-written Programs
- Stata Website

The About Stata menu item is the last item on Windows computers. With the exception of Search..., Stata Command..., Stata Website, and About Stata, all of these menu items open a Viewer window with the specific information requested.

On Macintosh computers, the Search field at the top of the Help menu is a Macintosh standard feature in all programs. Type a word in this field that is part of a menu item and a list of menu items appears; roll the mouse cursor over the menu items and it will indicate where that menu item is located.

The Advice, Contents, What's New, and News menu items offer helpful information that new users, in particular, may find useful. The News menu has current information about upcoming classes, etc. The Graph Editor brings up information specific to using the Graph Editor. What's New brings up information about the current version installed of both the Stata executable and ado-files. The SJ and User-written Programs item is to search and download files associated with the *Stata Journal* and older Stata technical bulletins. In addition, other locations are available that can be searched for user-developed programs for installation. This Viewer window also can list, search, and update previously installed program packages that you have downloaded.

The last menu item, Stata Website, has three submenus: Main Website, User Support, and Frequently Asked Questions on Macintosh computers. On Windows and Unix computers, it includes The Stata Blog, The *Stata Journal*, and Stata Press. The Main Website will automatically load Stata's main Web site in your default browser. User Support loads Stata's user support Web site. The Frequently Asked Questions loads Stata's Web page of frequently asked questions.

Stata's commitment to user support is evident. Internet access dramatically increases your access to Stata support, additional files and programs, and the ability to take Netcourses if you wish. Stata technical support is very responsive answering both simple questions about the Stata program and complex questions about statistics. They are easily reached via email and usually respond within a few days. All updates are free with a perpetual license—no annual fee or payments for updates. These updates are not insignificant and they are available quite often as Stata personnel routinely update the program and make these changes available to users.

As you begin to use the program, many of the dialog boxes used to implement various commands have common elements that appear at the bottom of these windows (Figure 1.9). The question mark, when selected, opens a Viewer window with information on using the selected command. The R button resets the dialog box clearing previous entries and the copy button does just that, copies the command to the clipboard. The OK and Submit buttons execute the command with the OK button closing the dialog box with execution, whereas Submit leaves the dialog box open. The Cancel closes the dialog box without executing the command.

**Figure 1.9** The bottom of many dialog boxes have similar elements with a question mark, R, and copy buttons on the lower left and OK, Cancel, and Submit buttons on the lower right.

## Conclusion

This first chapter was to give a quick overview of the main features and operation of Stata. I would urge you to read the *Getting Started with Stata* book for your particular operating system if you haven't already done so. In addition, I would recommend reading through the *User's Guide*. Both of these volumes will give a much better feel for how Stata operates with many examples and illustrations.

# 2
# Data Entry

Surprisingly, data entry and manipulation can be one of the most time-consuming parts of statistical analyses. In some cases, the actual statistical analyses may be inconsequential compared to the work of getting data into the program in the right format. This can be particularly problematic if you are getting data you didn't create. A colleague or official government source may give you data in a form that must be manipulated in some significant way prior to analysis. Stata offers a wealth of commands for just such purposes that can make quick work of the most intractable dataset. In fact, there is a reference manual devoted to the subject, called *Data Management** from Stata Corporation.

Data in Stata is handled in a spreadsheet format with columns as variables or identifiers and rows as observations. The easiest way to enter data directly into Stata is with the Data Editor. In Table 2.1 is a small piece of data. Try entering it in the Data Editor. If you double click on the gray cells at the top of an empty column, a dialog box appears letting you name the variable and set some parameters associated with it.

You can enter data and the Data Editor will give the column a generic name such as var1 (Figure 2.1). The name must be 1–32 characters long and begin with a letter or underscore. In addition, it cannot have any spaces. The column can be labeled with a word or phrase up to 80 characters long and can be used to give a fuller explanation of what that variable is.

The Properties Region includes information about the file and data and how the data will appear. The %9.0g is a format command. The % indicates it is a format. The 9.0g indicates the field width is nine characters wide and the .0 tells Stata to display as many decimals as were entered. The g indicates that the format is a general format. If you can't see all of your entry in the Data Editor, enter a larger number, such as 15.0g, to increase the width of the variable column. There

* Stata Press, 2011. *Data Management.* College Station: Texas.

**Table 2.1** Onion variety trial yields (15-ft plots)

| VARIETY | REPLICATION | YIELD |
|---|---|---|
| 1 | 1 | 95.1 |
| 1 | 2 | 107.4 |
| 1 | 3 | 97.7 |
| 1 | 4 | 101.3 |
| 2 | 1 | 116.5 |
| 2 | 2 | 97.8 |
| 2 | 3 | 103.6 |
| 2 | 4 | 101.4 |
| 3 | 1 | 108.6 |
| 3 | 2 | 98.6 |
| 3 | 3 | 82.5 |
| 3 | 4 | 90.9 |
| 4 | 1 | 122.5 |
| 4 | 2 | 120.9 |
| 4 | 3 | 99.5 |
| 4 | 4 | 99.2 |
| 5 | 1 | 86.8 |
| 5 | 2 | 105.2 |
| 5 | 3 | 98.6 |
| 5 | 4 | 113.9 |

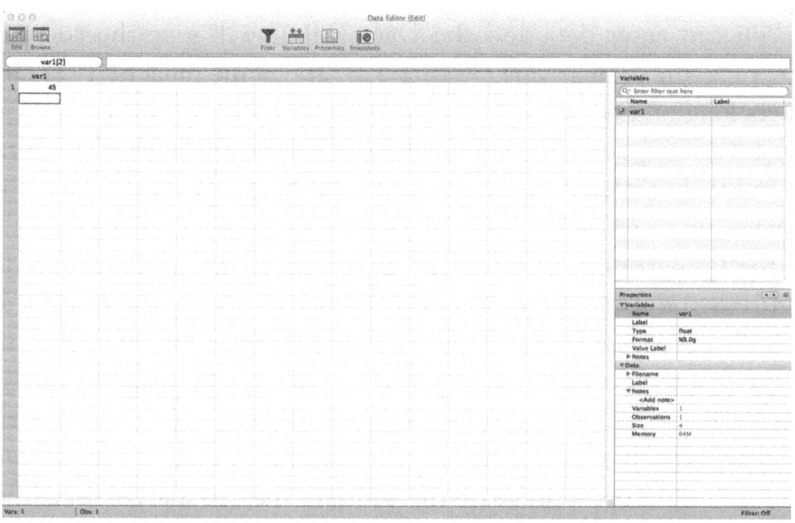

**Figure 2.1** Data Editor with one variable (var1) and one data point.

are many formats available for a variety of situations including formats for text, numbers, dates, and time.

A Value Label can be used to substitute a meaningful label for a number in the dataset. For example, your dataset may use a 1 for male and 2 for female. Someone reading that variable (even if labeled *sex*) may not know how the numbers are used. A value label can solve this problem. Select the button in the Value Label and another dialog appears. Select the Create Label button and give this value label a name (Figure 2.2). At this point, you can begin entering the numbers from this variable and giving them a value label. Once named, the label names can be defined for the numeric data. Data in the Data Editor will appear in one of three colors: black, red, or blue. Numeric values will be black,

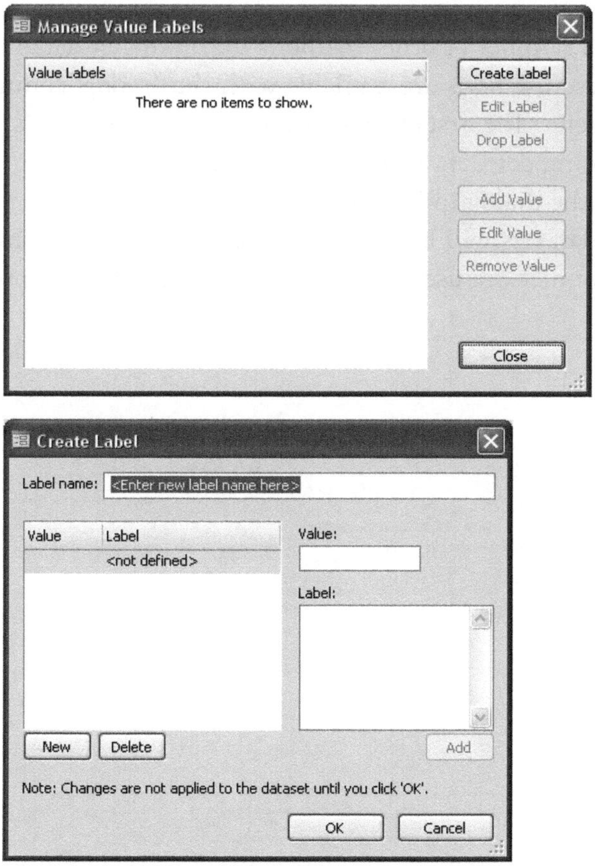


**Figure 2.2** Manage Value Labels window for creating value labels and the Create Label dialog after the Create Label button has been selected on a Windows computer.

text will be red, and value labels will appear blue. Stata can only use numeric and value labels (actually it still is a numeric value to Stata) variables for analyses. Text variables are used just as identifiers.

There is always the possibility of data being entered incorrectly. In fact, we have tried to reduce this from occurring by using data entry computers in the field (e.g., Ipad, etc.). This has become possible with the reduction in size and price of many such devices. In addition, more and more laboratory equipment saves collected data that can be imported to your computer further eliminating the possibility of data entry errors. Stata also helps by having a command, called **`assert`**, that allows you to check for data entry errors.

## Importing Data

Stata has a number of other methods for inputting data into the program. These methods are available with the Import command under the File menu. The first of these is Excel spreadsheet (*.xls, *xlsx). This is used to import data from Microsoft Excel®. Selecting this option brings up the dialog in Figure 2.3.

Many datasets will be tab or comma delimited. This means either tabs or commas are used to separate the data into columns. The first

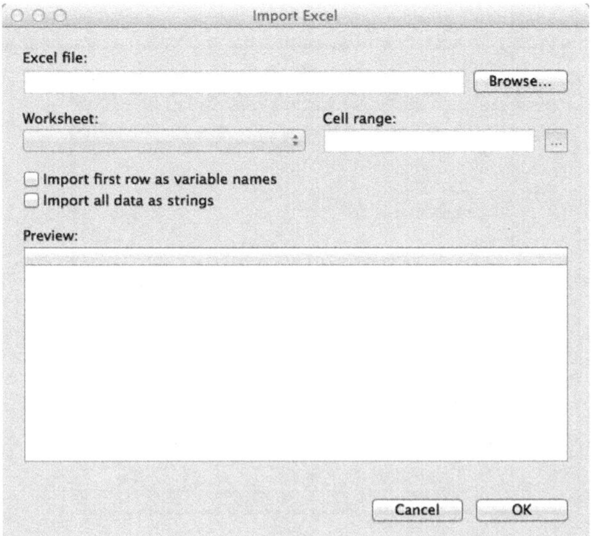


**Figure 2.3** Excel importing dialog for selecting an Excel workbook on a Macintosh computer.

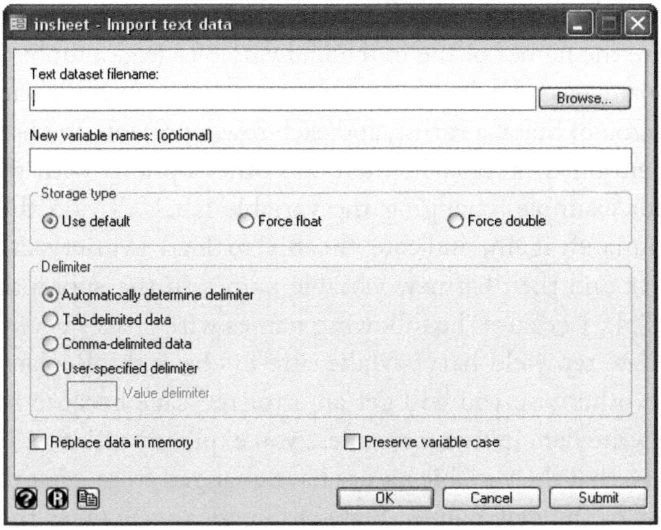


**Figure 2.4** Importing data created in a spreadsheet program on a Windows computer.

row can be the variable labels, but the remainder of the spreadsheet must be the data only. If you have a spreadsheet with header information, such as experiment name, date, etc., this method won't work. Selecting Text data created by a spreadsheet allows the importation of such files (Figure 2.4).

The file name can be typed in the 'Text dataset filename: field or click the Browse…' button to open a standard file dialog, find the file, click open, and the pathname is entered into the field. Remember, if you type the file name yourself, you will have to type the entire pathname. This can get quite convoluted if the file is buried several subdirectories deep. One way to avoid this is to change the working directory. Then, all you have to do is type the actual filename with its file extension if it is not the extension .raw. There are several other options available with this command including changing the storage type, variable labels, and the delimiter.

Try this function. Select the Text data created with a spreadsheet under File/Import. Then navigate to the file Variety 2000 Test Data .txt. There are several different file extensions this importing method supports including comma separated values (.csv), text files (.txt), and raw files (.raw). All of these are types of text files. This is a file that was originally created in Excel and saved as a text file. Once you have loaded it into Stata, you can view the data by selecting the Data Editor button in the Main window. You will notice across the top of the Data Editor are the names of the individual variables (e.g., number, variety, harvest date, etc.). Each column then represents a specific variable or, in the jargon of Stata, a varlist, and each row represents an observation.

As mentioned previously, there are other options with this command, for example, changing the variable labels. To do this, select this command again, indicate the file to load (Variety 2000 Test Data .txt), and then list new variable names in the appropriate field (Figure 2.4). Let's use the following names with spaces between each (no var date rep yield harv). Make sure to check the Replace data in memory; otherwise you will get an error message because Stata will not overwrite data in memory unless you explicitly tell it to. Now you will notice that the variable names have changed from what they were originally to the new names. Stata automatically changes the case of variable names to lower case, but you can force Stata to maintain the case by checking the Preserve variable case checkbox.

Another option with this function is the selection of storage type. Generally, you would leave this as Use Default. This lets Stata determine the appropriate storage type. When you first viewed the Data Editor, you would have noticed a couple of columns were in red indicating they were text or string variables. This is because Stata has interpreted these variables as strings. Numeric data (black) can be forced to a specific data type with this command, either as a float or long variable. These data types are used for numbers with many decimal places (more precision) and require more computer memory for each data point. In general, it is best to let Stata determine the appropriate data type.

This command also can be set to use specific delimiters, i.e., what character is used to separate the variables. Generally, it is best to let Stata determine this, but you can select a specific delimiter. This may be useful in a case where more than one delimiter character is in a dataset, such as commas and tabs, and the tabs are the delimiters you wish to use. The commas are just part of numbers (e.g., 9,999).

Finally, at the bottom of this dialog window are several icons (Figure 2.4). The question mark icon will open a Viewer window with information on using this particular command. The R will reset the dialog to an empty condition clearing all the fields. The final icon looks like two pages and copies this command to the clipboard. You can then paste the command into an editor, word processor, or the Command region of the Main window. This can be helpful in learning the command line structure. You can change different parameters in the dialog window and see how the command line is changed.

There are three buttons in the lower right of the dialog window that act as they would in most GUI (graphical user interface) programs. The Submit button executes the command, but leaves the dialog window open. The Cancel does just that, cancels the command and closes the dialog. The OK button executes the command and closes the dialog.

This may be a good time to talk about computer file types. Programs have specific file types that they use. For example, Microsoft Word or Excel have specific file types they use with specific extensions, .docx or .xlsx files. There are other file types that are generic that are meant to be shared between programs. Files of this type can be text or data types with identifying extensions, such as .txt or .csv. These latter file types are set up in a standard fashion so that many programs can interpret them. Stata also can interpret many of these files if they conform to specific layouts, such as a spreadsheet format of columns and rows. This does not mean Stata is incapable of reading files that don't conform to this layout as we shall see shortly.

All of Stata's commands available as specific menu items can be invoked by typing the command in the Command area of the Main window. In the case of Text data created by a spreadsheet, this command can be invoked by typing **`insheet using`** followed by the filename. The filename must contain the entire pathname (all the subdirectories), which are entered in a specific format based on your operating system. In Windows, subdirectories are separated with the back slash (\), and with Unix and Macintosh, the forward slash (/) is used. Stata, however, is smart enough to recognize either back slash (\) or forward slash (/) on all operating systems. The entire pathname, however, is not required if you have changed the working directory (Change Working Directory…) under the File menu to the directory where the data file is stored. This can make using this command much easier because only the filename is now required to be entered, not the entire pathname. It is a good practice to change the working directory each time you start Stata to the directory where your working files are stored. With this book, the example files will be stored in the Data folder and available to you.

Stata uses a specific language syntax to invoke a command. This syntax is common across all of Stata's commands and is explained in the help files available through the Viewer window. Using the **insheet using** command as an example, this command looks like this in its help window:

**insheet** [*varlist*] **using** *filename* [, *options*]

Items without brackets are required so the command at the very least would include **insheet using** *filename*. Items in brackets are optional and may or may not be used. In the above command, the *varlist* changes the variable names in the imported file. Remember, we changed the variable names from number, variety, etc. to no, var, etc. in the dialog window. The comma is required if any options are to be used. These options include such things as changing the file delimiter and data type, and maintaining name case, etc. For a more detailed explanation of this command, open a Viewer window and type **help insheet** in the Command field at the top of the window. This will bring up the help file for this command. At the top right of the window is a drop-down menu Dialog where the **insheet** dialog can be opened. This can be helpful if, for example, while looking through the help files, you find a command you are interested in, but are unsure how the command works. Select the drop-down item and the dialog window will appear, which you can then fill out. The help information listed in the Viewer window includes under which menu this particular command is located. It will be listed under the Menu heading.

Some commands will have the first letter or two underlined; this means that you can abbreviate that command by typing just that letter or two. Look at the example below the **describe** command. Notice how the **d** is underlined, which means this command could be invoked by typing just the **d** in the command window. Some commands require the entire command be typed. This is reserved for commands that will change something that can't be undone. This helps protect you from accidentally and irretrievably changing your data. For example, the command **generate** that generates a new variable can be entered with just **g**, while **replace** requires the entire word be entered to avoid inadvertently replacing important data. It is still possible to make mistakes, but this should help some. To see how abbreviated commands work, try the below command with just **d**.

**describe** [*varlist*] [, *memory_options*]

Another method for importing data into Stata is to use a data dictionary. This involves two files: (1) the data file and (2) a data dictionary file that tells Stata how to interpret and import the data. Often data files are not just the data, but rather have additional information about the experiment. This may be several rows of information at the beginning of a file before the actual data. An example of this is shown in Figure 2.5.

```
Vidalia Onion and Vegetable Research Center 2001 Onion Variety Trial

9/25/00 Sowed beds
11/27/00 onions transplanted

Plant Beds - Fumigated 8/15/00 with 63 gal. 42% metam sodium per acre
      Field Production - Transplanted 11/27/00
9/12/00 1 ton dolomitic lime per acre

Plantbed Fertility                      Fertility
9/19/00 800 lbs 5-10-15 (9% sulfur)     11/9/00 400 lbs. 5-10-15 (9% sulfur)

9/26/00 150 lbs. 18-46-0 peracre        12/20/00 150 lbs. 18-46-0
10/26/00 200 lbs CaNO3 per acre         1/2/01 200 lbs. 6-12-18 (5% sulfur)
11/9/00 100 lbs CaNO3 per acre          1/16/01 200 lbs. 6-12-18 (5% sulfur)

Total = 113-149-120 (72 lbs. Sulfur)    2/7/01 200 lbs. 15.5-0-0
                                        2/20/01 200 lbs. 15.5-0-0
                                        Total = 133-157-132 (56 lbs. Sulfur)


Replication      Variety       Date         Field Yield
      1             1        5/10/01           78.2
      1             1        5/10/01           65.7
      2             1        5/10/01           82.6
      2             1        5/10/01           61.1
      3             1        5/10/01           78.2
      3             1        5/10/01           61.7
      4             1        5/10/01           78.4
      4             1        5/10/01           52.5
      1             2         5/3/01           44.8
      1             2         5/3/01           46.5
      1             2         5/3/01           48.3
      2             2         5/3/01           44.7
      2             2         5/3/01           45.3
      2             2         5/3/01           47.3
      3             2         5/3/01           46.9
      3             2         5/3/01           47.1
      3             2         5/3/01           48.9
```

**Figure 2.5** Example text file with information about the experiment at the top and a segment of data below.

```
dictionary {
_firstlineoffile(22)
_lines(1)
        int rep
        int var
        str8 date
        float yield
        }
```

**Figure 2.6** Example of a data dictionary used to import data.

This is where a data dictionary can be helpful in inputting such data. Only the data with the column labels should become part of the data file. To do this, you must create a data dictionary file, which is just a plain text file set up in such a way that Stata can use this dictionary to determine how the data should be input. The Do-File Editor is a good place to create this file, although any text editor or word processor can be used.

Figure 2.6 has the data dictionary created to import the data from the file shown in Figure 2.5.

The data dictionary must have the word dictionary in the first line with an open brace ({). The first line after that tells Stata that data should be imported starting at line 22. The next line indicates that each observation is on one line. This is really not necessary in this case because Stata can figure this out. It does imply, however, that data can be imported with a single observation that is contained on more than one line. The next four lines indicate the data type and variable name for each variable. **int** is for an integer type, which is a number without any decimals; **str8** is for a string or text type that is eight characters or less in length. **float** indicates that data type is a floating number, i.e., it has a decimal value. These explanations are somewhat simplistic and Stata's Help files and manuals have more detailed information about data types. Dictionaries should be saved with a .dct extension so Stata will recognize them.

Figure 2.7 shows the input dialog window. This window is selected under the File menu, under the Import submenu. Select the submenu item: Text data in a fixed format with a dictionary. Select the Browse… button for the dictionary filename and load the Infiledict .dct file from the Data Files folder. Then do the same to load the text

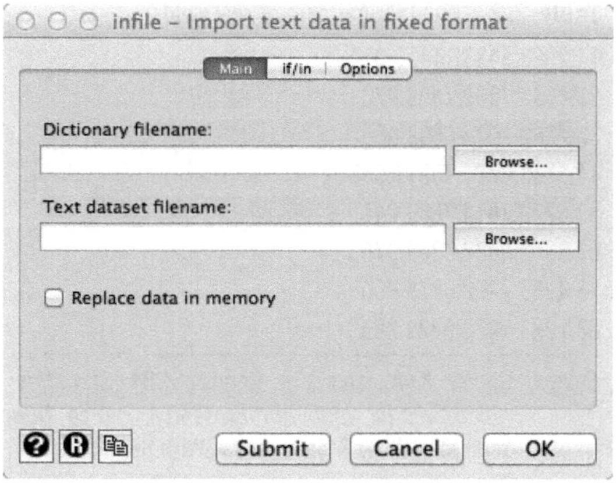


**Figure 2.7** The dialog box for importing from a text file with a dictionary.

dataset filename, called Variety 2001 .raw. The dictionary file is used by Stata to interpret how the data file should be loaded.

This import method can be entered in the Command area of the Main window with either of these commands:

```
infile using "/Users/georgeboyhan/Documents/Books/
Stata Book Files/Data Files/Infiledict.dct",
using ("/Users/georgeboyhan/Documents/Books/Stata Book
Files/Data Files/Variety 2001.raw")

infile using Infiledict.dct, using ("Variety 2001.raw")
```

The first instance of the command is what is echoed to the Results area of the Main window when the command is entered from the dialog window (Figure 2.7). The second instance is what I typed after changing the working directory to where the files are stored. You can see how the pathname is no longer needed. This can make things easier if you are using the Command area of the Main window.

The next command to look at for importing data is **infix**, which is used to import fixed formatted data. That is data that have a fixed width for each data point. Look at the data in Table 2.2; although it doesn't look like it, it is in a fixed format that Stata can easily import. The table fragment is from page 77 of *Statistical Procedures for Agricultural Research* (Gomez and Gomez, 1984) and was nicely formatted; however, upon scanning into a computer the formatting was lost. The first column, which consists of the numbers 1–8, are the

**Table 2.2** Fixed format data of rice yields

| |
|---|
| 14.2523.5483.11410.914 |
| 23.4632.7202.7898.972 |
| 33.2282.7972.8608.885 |
| 44.1533.6723.73811.563 |
| 53.6722.7812.7889.241 |
| 63.3372.8032.9369.076 |
| 73.4983.7252.6279.850 |
| 83.2223.1422.9229.286 |

*Source:* Gomez, K. A., and A. A. Gomez. 1984. *Statistical procedures for agricultural research,* 2nd ed. New York: John Wiley & Sons, p. 77. With permission.

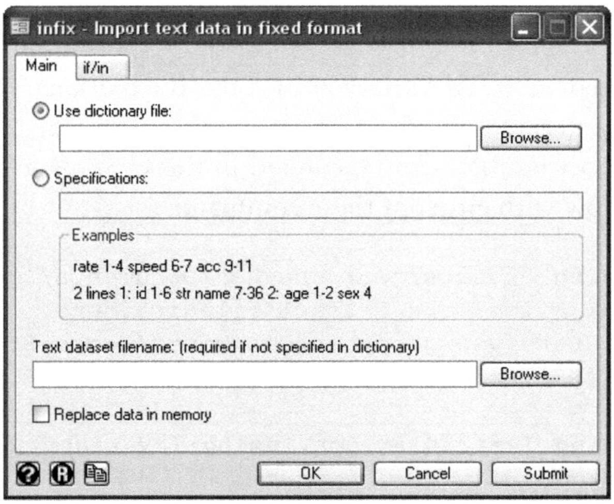


**Figure 2.8** Importing data in a fixed format dialog box on a Windows computer. Note that the dictionary can be used with this command as well.

variety numbers. The next three columns consist of four characters each and represent each of three replications. For example, in the first row the grain yields are 4.252, 3.548, and 3.114. The last column is the total for the three replications, which in the first row is 10.914. Knowing this makes it easy to import these data into Stata.

To import these data, select the Text data in fixed format under the Import submenu, under the File menu (Figure 2.8).

Select the Specifications: button and enter the following: var 1 rep1 2-6 rep2 7-11 rep3 12-16 total 17-22. Then select the Browse… button and select the riceyield.txt file. Stata defaults to a .raw file extension, thus this file may appear dimmed until you enable the .txt file extension. If you entered the specifications correctly, you should have a file with eight observations and five variables.

To use this command from the Command area of the Main window, type in

```
infix var 1 rep1 2-6 rep2 7-11 rep3 12-16 total 17-22
using riceyield.txt
```

You may have noticed that I didn't type nearly as much as was echoed to the Results window when using the dialog window. This is because I changed the working directory to where the data files are stored so the pathname does not have to be typed.

There are occasions when not all of the data for an experiment are in the same file. For example, yield data may have been collected at different times or even over several years and each time the data were collected they were entered into a different file. Stata has commands that make merging data relatively easy.

There are three files available online* we will use to illustrate one method of merging data. The data are from a watermelon variety trial that was harvested on three separate days, thus the three files. The files contain a variable called *entry*, which denotes the plot number and five columns of variables, which are the weights of individual fruit. What we want to do is append two files onto the end of the third. Stata uses the term *master* to describe the file in memory and the files that will be appended to the master as the *using datasets.*

Open the dataset labeled water71503.dta; this will be the master dataset. To this dataset we will append the using datasets of water71603.dta and water72103.dta. To do this, under the Data menu, select

```
Combine datasets > Append datasets
```

This will bring up the append—Append datasets dialog window (Figure 2.9). Use the Browse… button as before and find the second file water71603.dta. Then select the Select an additional file button, which then allows you to select an additional file, water72103.dta. Leave the other options as they are and click the OK button. This will merge the using datasets water71603.dta and water72103.dta with the

* Files available online at http://www.crcpress.com/products/isbn/9781466585850.

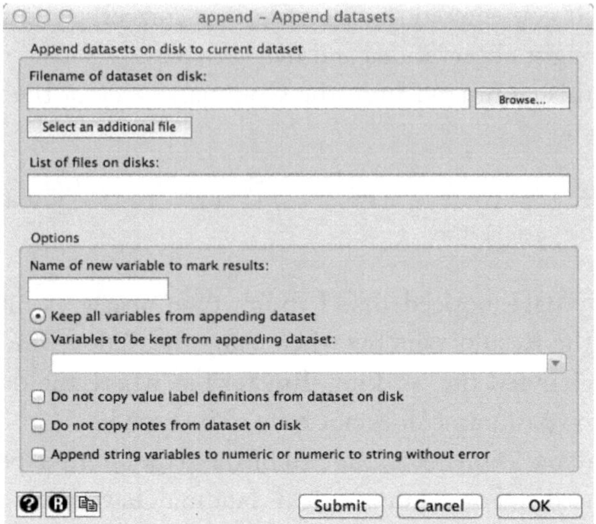


**Figure 2.9** Append dialog box for appending a file on disk to the one in memory.

master dataset water71503.dta by appending the using dataset to the bottom of the master dataset.

If you looked at the master dataset before appending the other files, you would have seen 82 observations; after appending, there are 412 observations.

The command for this type of document merge is

```
append using filename [filename...] [, options]
```

Remember the master file should have already been loaded into memory before appending the using datasets with this command. Again more detailed information about the options are available from the Help file.

Another method Stata has to merge files is to merge them side-by-side. Look at Table 2.3, which illustrates this type of document merge.

In this example, a dataset was created when data on seedstems (flowering) and doubles (doubled bulbs), which are undesirable characteristics in onions, were collected from an onion variety trial. In addition, a stand count was made of all the plots. Later, the plots were harvested and the yield data were collected, which were entered into a separate dataset. At some point, it was decided to merge these datasets in a side-by-side fashion to do additional analyses.

**Table 2.3** Illustration of merging two files

| MASTER DATASET (IN MEMORY) | | | | | USING DATASET (ON DISK) | | |
|---|---|---|---|---|---|---|---|
| REP | ENTRY | PLANT COUNT | SEED STEMS | DOUBLES | ENTRY | REP | FIELD YIELDS |
| 1 | 1 | 21 | 0 | 4 | 1 | 1 | 156.6 |
| 2 | 2 | 33 | 1 | 1 | 2 | 2 | 92.0 |
| 3 | 3 | 41 | 0 | 4 | 3 | 3 | 117.4 |
| 4 | 4 | 5 | 0 | 1 | 4 | 4 | 109.2 |

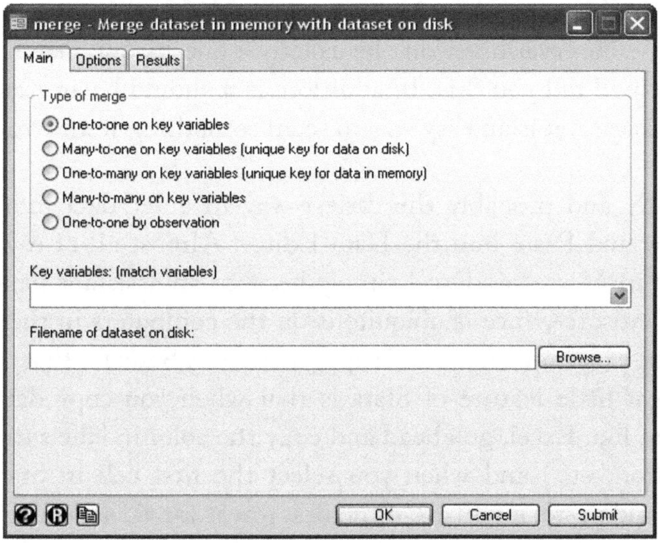


**Figure 2.10** Windows computer dialog for merging datasets side-by-side.

To accomplish this, the first dataset, onioncount03.dta, should be opened. This will be the master dataset. Then select Merge two datasets (Figure 2.10).

```
Data > Combine datasets > Merge two datasets
```

Select the onionyield03.dta file as the file to be merged into the master file.

There are several options for the type of merge that can be accomplished. In this case, the appropriate selection is One-to-one by observation. The other three options require a unique identifier for each observation in one or both files.

If all went well, when you open the Data Editor you should have a dataset with the five variables from the first dataset (rep, entry, plantcount, seedstems, and doubles) along with the fieldyield variable from the second dataset. Stata recognized the rep and entry variables in both datasets, thus, the merge was done across these variables and they only appear once in the new file. In addition, there is another variable called _merge, which in this case is a list of 3s. In a merge, Stata will create this new variable and it will indicate where the specific observation came from. For example, if an observation is only in the master file, a 1 will appear in this column. If the observation is only from the using file, a 2 will appear, and if from both, a 3. In some cases, the observations won't be exactly a one-to-one match and this variable will tell you this. In addition, if it should be an exact match as in this case, it is an easy way to see if something went wrong in the merge.

Finally and probably the easiest way to enter data into Stata is to Copy and Paste into the Data Editor. Almost all of my data are entered in Microsoft Excel simply because an assistant handles this and Microsoft Office is ubiquitous on the computers in the office. It would be a bit impractical and expensive to buy Stata for that computer.

A neat little feature of Stata is that when you copy data from a program like Excel, go ahead and copy the column labels (treatment, replication, etc.) and when you select the first cell in Stata's Data Editor and paste, it will ask if the first row is for variable labels. It will even make adjustments to the names if there are any conflicts (Stata requires that all variable labels should have unique values).

There are several other methods of importing data into Stata and I will leave it to you to explore them if necessary. In addition, the import examples shown here can be even more flexible with their capabilities especially when using a data dictionary.

### Manipulating Data and Formats

Stata can be useful even before you begin an experiment by generating random number tables that are organized for your specific experiment. For example, if you have an experiment that is going to be a randomized complete block design (RCBD) with 12 treatments and 4 replications, you would want treatments 1–12 randomized within each of 4 blocks (replications). The generated randomization then can be taken to the field, greenhouse, etc. to install the experiment.

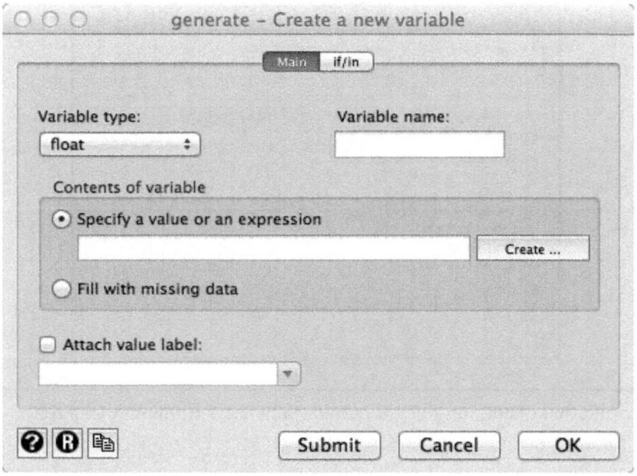


**Figure 2.11** Dialog for creating new variables.

To develop this randomization, start with an empty Data Editor (no dataset in memory). In this case, we will need an empty dataset of 48 missing values. There is no dialog window to do this; it must be done in the Command area of the Main window. Type the following in the Command region:

```
set obs 48
```

This will create 48 blank rows in the Data Editor. Then generate a new variable with random numbers. To do this, select the Create new variable submenu (Figure 2.11).

```
Data > Create or change variables > Create new variable
```

Enter a name for this variable in the Variable name: and then enter **runiform()**in the Specify a value or an expression field. The **runiform()** function can be selected under the Create… button from the Random numbers category under Functions (Figure 2.12). The Create… button to the right of the Specify a value or an expression field looks a little like a calculator and that is its use. A variety of functions will appear in the right list when different categories in the left list are selected, as well as math functions and logical operators. This dialog is used by many different commands.

To set this from the Command window, type

```
generate x = runiform()
```

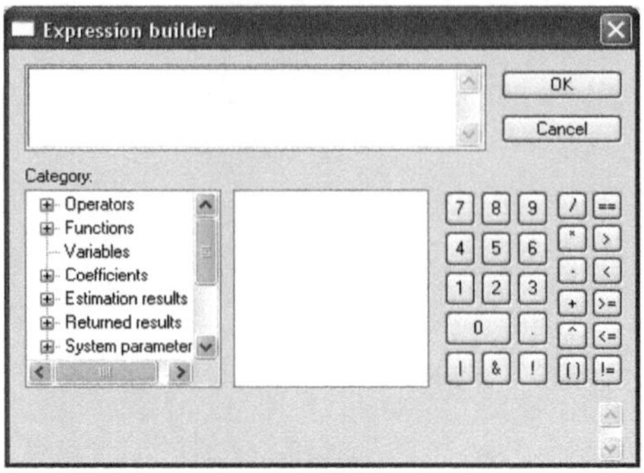


**Figure 2.12** Function dialog for filling a new variable on a Windows computer.

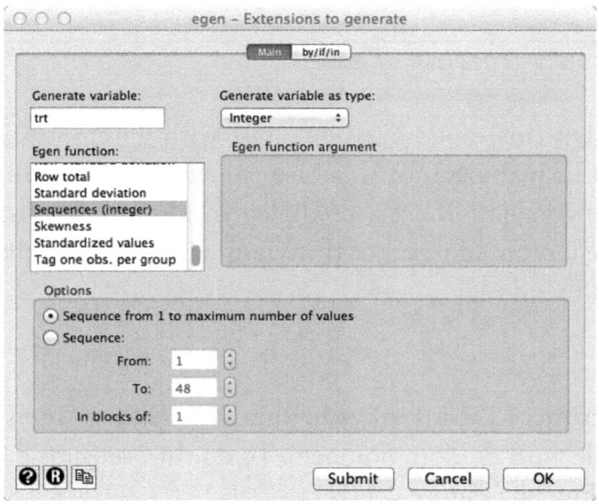


**Figure 2.13** Extensions to generating new variables with the variable trt entered, Sequences (integer) selected, and the variable type, Integer.

Remember, as mentioned before with some commands, only part of the name needs to be typed. In this case, typing **g** is sufficient. The **runiform()** function generates a uniformly distributed set of random numbers on the interval 0–1. At any point in this process, you can select the Data Editor button at the top of the Main window or type **edit** in the Command window to see how the dataset is changed with each command.

At this point, we want to generate a new variable with four groups with numbers 1–12. This represents the 12 treatments in each replication group. To do this, select the submenu Create new variable (extended) (Figure 2.13).

```
Data > Create or change data > Create new variable
(extended)
```

In this dialog, enter a name in the Generate variable: field, for example trt. From the Egen function: list, select Sequences (integer), which will change the option portion of the dialog window. In the Options select the Sequence: button and fill in from 1 to 12 in blocks of 1. It may seem that the In blocks of: field should have a 4, but 1 is correct. You may wish to try it both ways just to see how the numbers are generated. From the Command area of the Main window, type in

```
egen trt = seq(), from (1) to(12) block(1)
```

Next, select the same dialog window, enter a new variable, rep, and enter the sequence from 1 to 4 in blocks of 12. The Command entry is

```
egen rep = seq(), from (1) to(4) block(12)
```

Next, you will want to sort your data using the random variable in groups of 12. To do this, select the Ascending sort submenu (Figure 2.14).

```
Data > Sort > Ascending sort
```

Select the random variable in the Variables: field and check the box for Restrict sort of observations and enter from 1 to 12. This will randomly sort the first 12 treatments in replication 1. Do this three more

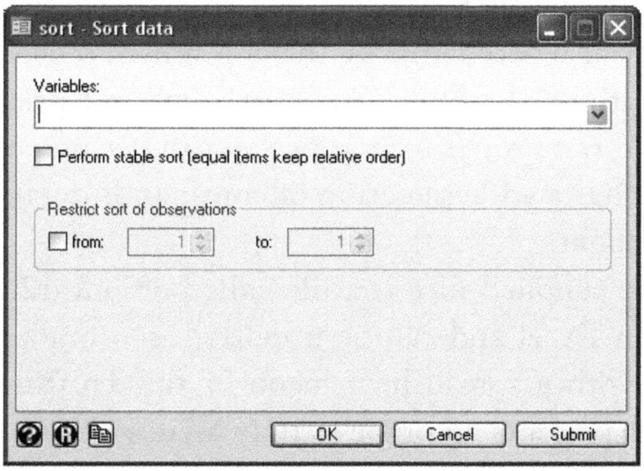


**Figure 2.14** Sort dialog on a Windows computer.

times selecting observations 13–24, then 25–36, and, finally, 37–48. To do this from the Command window, enter

```
sort x in 1/12
sort x in 13/24
sort x in 25/36
sort x in 37/48
```

You can now drop the random variable by selecting the Variables manager submenu, then select the random variable and hit backspace.

```
Data > Variables manager
```

To do this in the Command window, type

```
drop random
```

You now have a list of 12 treatments randomized within each of 4 replications. This can be printed out and taken to the field or used to label and organize stakes, etc.

Once you have your data in a Stata file there are many types of changes and additions that can be accomplished before conducting your analyses. Often the rows represent specific treatments and the columns are observations or replications. Many textbook examples present data in this format, which is different than required in Stata. If you are trying to learn a new method and scan in a table from a textbook to try it in Stata, you may find you have to rearrange the data prior to analysis.

An example of this was presented in the last section with the rice variety trial where three of the columns represented the individual replications. In Stata, you can copy and paste the data to rearrange it to the proper format or you can use one of Stata's commands. Stata has a wide selection of commands dealing solely with data management.

Among the supplied files is a file called onionvar2003.txt, which was created in Excel and can be imported into Stata with the Text data created with a spreadsheet menu or the **insheet** command. This dataset represents an onion variety trial with 30 varieties and 4 replications (variety 27 had crop failure) with the rows or observations of the varieties and columns as replications.

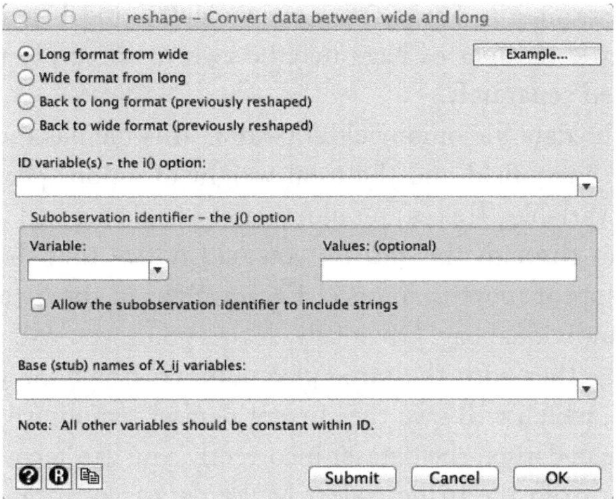


**Figure 2.15** Dialog box to convert a dataset from a wide-to-long or long-to-wide format.

In the vernacular of Stata, this dataset is in a wide format and it needs to be in a long format for the type of analysis we want to make. This can be accomplished relatively easily by selecting the Convert data between wide and long submenu (Figure 2.15).

```
Data > Create or change data > Other variable-transfor-
mation commands > Convert data between wide and long
```

In order for this to work, the replications have to have a stub name with a unique ending. In this case, the stub is *rep* with the unique endings 1–4. In the ID variable(s) – the i() option: select the entry variable. In the Subobservation identifier variable – the j() option: enter num or you can use any variable name you wish, in the Variable: field, since this command is going to create the variable. Finally, in the Base (stub) names of X_ij variables: enter rep and then click the OK button. The num variable is now your replication identifier and the onion yield variable is labeled *rep*. You may wish to change these names to reflect what they are, perhaps rep and yield, respectively, for the replication and onion yield.

Oftentimes you may have more than one measurement for each experimental unit. For example, in a watermelon trial, each plot may have a couple hundred pounds of watermelon that cannot be weighed all at once. This results in multiple weights for each plot.

Another example is onions that are collected and bagged for each plot, and the number of bags may be two to three per plot, which are weighed separately.

Open the data file onionyield2002.dta. This file has the plot number in the entry field and the total weight of onions per plot in the weightlbs variable. This is an onion variety trial that had 31 varieties. If you scan through the entries, you will notice that the same plot numbers appear more than once. Each weight in the dataset is actually of an individual bag. Obviously, at this point, you will want to add weights together with the same plot number. Enter the **preserve** command, which will save the current dataset temporarily. This way, if you have collapsed the dataset incorrectly, you can recover the data. To do this from the menu, select Make dataset of means, medians, etc.

```
Data > Create or change variables > Other variable-
transformation commands > Make dataset of means, medi-
ans, etc.
```

In the dialog window, select Sum on the drop-down menu for number 1 under Statistic and enter weightlbs in the Variables field (Figure 2.16). Then select the Options tab at the top of the window and, in the Grouping variables: field, enter entry and click OK. This

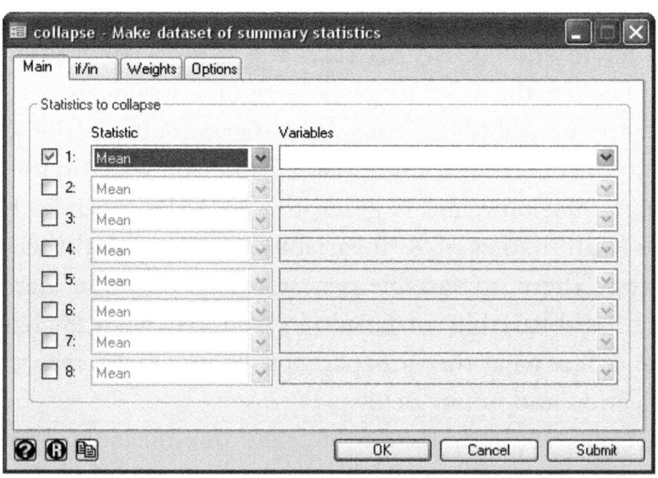


**Figure 2.16** Collapse dialog for making dataset of summary statistics on a Windows computer.

command collapses the dataset adding all the weights with the same entry number.

To do this from the command window, enter

```
collapse (sum) weightlbs, by(entry)
```

The **(sum)** indicates that the command should add the weights weightlbs by the entries entry. If you look at the Help screen for this command in the Viewer window, you will see that a dataset can be collapsed by the mean, which is the default if nothing is specified in the command, but you also have several other alternatives, such as median, standard deviation, etc. If the collapsed dataset isn't correct or you made a mistake, use the **restore** command to restore the original dataset. It gives you a great deal of versatility in handling a dataset.

Once the file is collapsed, you will need to create variables to represent the different parameters in the model. If you have the collapsed file open from the previous paragraph, the next step is to identify the treatments (varieties in this case) and the replications for an RCBD. If not, open the file onionyield2002collapsed.dta, which is the collapsed file from the previous paragraph. Most field experiments that set up are RCBD, which is probably the most common field design in agriculture (we will talk about them later). I will code such experiments with a three digit number where the first digit is the replication number and the next two are the treatment number. For example, with the plot number 403, the 4 represents the replication and the 03 the treatment.

To create a variable with the variety number, select the following menu:

```
Data > Create or change data > Create new variable
```

When the dialog appears, enter variety in the Variable name: field and, in the Specify a value or an expression, `enter` **mod** `(entry,100)`. Then press OK.

In our case, entering **mod** `(entry,100)` calculates the modulus of entry divided by 100. The modulus is the remainder from division, so, for example, dividing 403 by 100 has a remainder of 3. The modulus is the remainder.

To extract the replication number, you would select the same function dialog, enter rep for the new variable name and for the expression enter **int** (entry/100). This divides the entry by 100 and the **int** takes just the integer part of the number. With 403, divide by 100 and the integer portion is 4. In this case, we could have selected the menu item

```
Data > Create or change data > Change contents of
variable
```

Select entry under Variable:, which will be changed to the replication because we no longer need the entry information. I find it better to go ahead and create a new variable just in case.

Use these functions in the Command window:

```
generate variety = mod (entry,100)
generate rep = int (entry/100)
```

The first creates the variety variable extracted from entry and the second the replication (rep). At this point, this dataset would be ready for analysis.

Once you have entered or imported your data and arranged your variables for analysis, you may wish to add additional information to the dataset, such as the Label that was used from within the Data Editor. Such detailed information may not seem necessary when first working with a dataset, but over time you may forget what the data represented and how the experiment was arranged—dates, places, etc. Much of this detail information can be easily added to a dataset ensuring that, if you do have to come back to the dataset years later or if a colleague needs the information, it will still make sense.

There are two types of information that can be added to a dataset. The first are labels, which were covered earlier in the Data Entry section. Labels are short descriptions for variables or for the dataset as a whole. These are 80 characters or less in length. In addition, value labels can substitute a label for a variable number, such as a variety name or treatment name (see Data entry section). Labels appear when you use the **describe** command.

In addition to the labels, notes can be attached to a dataset or individual variables. To attach a label to the dataset, select the menu

```
Data > Data utilities > Label utilities > Label dataset
```

To attach a note to a variable, select the menu

```
Data > Variables Manager
```

To add a label to the dataset, in the Command window, type

```
label data: 'type your label'
```

To add a note to a variable, type

```
notes varname: your note
```

To list the notes from the Command window, type

```
notes
```

These labels and notes can be added or changed from the Main window by clicking the lock icon in the Properties region. This will unlock the variables properties where the label and notes can be accessed. Click the … button to add additional notes. These additions also can be made from within the Data Editor window in the Properties section of this window.

# 3
# DESCRIPTIVE STATISTICS

Before beginning, it might be a good idea to start a log of the session. This way all of your calculations are saved for future reference and use. I make a lot of mistakes when working with a dataset and sometimes I discover something new with the program. Having a log makes it easy to go back and see what was done and pull out information as I write reports and papers.

At this point, I will not be describing the menu location for a specific command, nor will there be illustrations of every command dialog window. Having read through Chapter 2, you should have a good idea how these work. If you want to know where a particular menu item is located, type **help** and the specific command in the Command region of the main menu or in a Viewer window and the Help file will list where a particular menu can be found.

After opening a dataset and manipulating the data and variables for analysis, one of the most useful commands is **describe**. This will give you an overview of what the dataset consists of. The command can be used on a dataset in memory or on disk. The formats for this command are

```
describe [varlist] [, memory_options]
describe [varlist] using filename [, file_options]
```

The **describe** command gives you information about how many observations are in the dataset, the number of variables, the storage type, display format, if there are value labels, and variable labels. If there is a dataset label, this also will be displayed.

If you have any familiarity with computers and programming, the storage type will be familiar to you. Different types of data take up different amounts of memory. For example, a byte is used for digits without decimal places, but has a small range of allowed values (–127 to 100), whereas variables stored as float have decimals and a relatively wide range of values (–1.70141173319*10^38 to 1.70141173319*10^38). Values stored as doubles can have an even wider range and greater precision (–8.9884656743*10^307 to 8.9884656743*10^307). A string variable is used for a string of characters. For more information on data types, type help data types in a Viewer window. Generally you don't have to be concerned with a variable's storage type unless memory is getting low, in which case switching a variable from a float type to byte type, for example, can save memory, assuming the variable can be stored as a byte. Even this level of knowledge is not that important because Stata has a command, **compress**, that can be used to reduce the size of the dataset in memory. With this command, Stata attempts to store variables in a smaller storage type if possible. The command is

```
compress [varlist]
```

It is helpful when working on a dataset to change the location of the working directory. This makes it easy to open or save files when entering commands in the Command window. You don't have to type the entire pathname. The command for changing the directory is

```
cd "directory name"
```

In statistics, one of the most important types of information you will want to look at and report is descriptive statistics, which includes measures of central tendency and dispersion. The most important measure of central tendency is the mean or average. This can be calculated and reported easily by Stata. Open the onionyield03.dta dataset, which we will use to demonstrate some of the descriptive statistics available in Stata.

One of the most useful commands for reporting means is the **tabstat** command.

```
tabstat varlist [if] [in] [weight] [, options]
```

The **tabstat** command is used to generate a variety of descriptive statistics. In the command listed above, you will see the terms **if**, **in**, and **weight**. These qualifiers are available with many Stata commands, so let's cover them now. The **if** qualifier allows you to select observations based on a condition. For example, you may be interested in only yields above a certain level or range. The **in** qualifier allows you to select a range of observations. For example, if you wish to look at the first 10 entries, type **tabstat** fieldyield **in** 1/10. Weights allow you to have another variable as a weight in computing the statistic of interest. For example, a specific measurement may have occurred several times for a treatment so that a frequency weight would be used. Stata also allows probability, analytic, and importance weights, which are discussed in more detail in the online Help files.

Open the onionyield03.dta for the following example. The **tabstat** command is most useful when used with a grouping command. In this case, the entry variable indicates the varieties that were in the experiment. The **tabstat** command can compute statistics by a specific variable (var, in this case) in a couple of different ways. Look at the following two commands:

```
tabstat fieldyield, statistics(mean) by(entry)
by entry, sort : tabstat fieldyield, statistics(mean)
```

Both of these commands do the same thing: list the means of fieldyield, which is computed for each entry. The first case is probably more useful because it lists the results as a simple table, whereas, in the second case, the headings are re-created each time a new mean is calculated. The first command will result in a data display that will be easier to cut and paste into another document as a table.

There are a large number of descriptive statistics that can be computed with the **tabstat** command. They include the mean, count, sum, maximum, minimum, range, standard deviation, variance, coefficient of variation, standard error of the mean, skewness, kurtosis, median, as well as a number of percentiles.

Although **tabstat** is capable of generating many descriptive statistics, it does not save any of these results. With some commands, Stata saves the results in variables that can be accessed for further calculations. One such command is **summarize**:

```
summarize [varlist] [if] [in] [weight] [, options]
```

**summarize** will calculate the number of observations, mean, standard deviation, minimum, and maximum. You can use this command to summarize with **by** or if unsorted use the **bysort** prefix command. The **by** is implemented as **by** *varlist*: **summarize** [*varlist*], where the first *varlist* is the grouping variable, in our case the varieties, while the second *varlist* is the variable to be computed, in our case the plot yield. There are a number of options available after the comma including **detail** that calculates several additional statistics. Below is an example of the output.

```
    Variable |       Obs        Mean    Std. Dev.       Min        Max
-------------+--------------------------------------------------------
  fieldyield |       120    93.82417    27.04157      41.7      214.8
```

**summarize** saves results in **r()**, which are specific variables for results of the most recent calculation. Commands that save in **r()** are called r-class commands. To see what these results are, type **return list** in the Command window immediately after entering a **summarize** command. The listed variables and their values will be shown in the Results window.

```
scalars:
                  r(N) =  120
              r(sum_w) =  120
               r(mean) =  93.82416639328002
                r(Var) =  731.2465641339238
                 r(sd) =  27.04157103671907
                r(min) =  41.70000076293945
                r(max) =  214.8000030517578
                r(sum) =  11258.8999671936
```

If you use the **by** prefix command, in this case the var variable, only the last calculation will have r() results listed—in this case, for variety Sapelo Sweet. As long as you do not enter another command that will overwrite these values, they are available for use in other calculations. These calculations also can be displayed immediately in the Results window with a new command **display**. Therefore, for example, you can calculate the standard error of the mean by taking the square root of the variance divided by the number of observations. To do this, enter the following command:

```
display sqrt(r(Var)/r(N))
```

The resulting value, 2.4685464, will be displayed in the Results window. The **sqrt()** function is one of many functions built into Stata that can be used in an immediate mode, such as we have done here. Typing **help functions** in a Viewer window will bring up a Help screen with a list of many types of functions.

Another example of using these saved values is to calculate confidence intervals. Confidence intervals indicate with a certain level of probability that the mean will fall in the confidence interval range. In this case, we know the mean (93.8) from the **summarize** command and we know the standard error of the mean (2.47) from our previous displayed calculation.

Confidence intervals are calculated from the mean plus or minus tabular *t* times the standard error of the mean ($\bar{y} \pm ts_{\bar{y}}$). Tabular *t* can be found in the back of most statistics texts in a *t* table. Stata has a function that can calculate this value for you, which is the **invttail**(n,p) where *n* is the degrees of freedom (one less than the number of observations) and *p* is the inverse probability. If we wish to see the 95% confidence limit, we would enter for *p*, 0.05. Actually, Stata only calculates the larger positive value with this function rather than the larger numeric value, so *p* should be 0.025 (half of 0.05) to be correct. Therefore, enter the following command:

```
display r(mean) " +/- " sqrt(r(Var)/r(N))*invttail(r(N)-1,0.025)
```

This command will display the mean plus or minus the confidence interval. To see what exactly the lower and upper confidence intervals are, enter

```
display r(mean - sqrt(r(Var)/r(N))*invttail(r(N)-1,0.025)
display r(mean + sqrt(r(Var)/r(N))*invttail(r(N)-1,0.025)
```

Of course, Stata has a command that will calculate the confidence interval for you:

```
ci [varlist] [if] [in] [weight] [, options]
```

The advantage of calculating this value yourself, as with other possible calculations, is that you can control the output as well as what combination of values you wish to display together.

### Output Formats

Stata has the ability to control the output for numbers, strings (text), and dates. Dataset variables are stored in a number of possible data types (storage formats) based on the data. For example, whole numbers may be stored as integers (`int`), which can be values from –32,767 to 32,740. There is a default display format for integers that is %8.0g. This format indicates there are 8 spaces for the number and the 0 indicates the number of decimals is left up to whatever the number is. In the case of integers, there are no decimal places. The g indicates it is a general format where the digits to display right of the decimal point are determined by the value, and if the value is too long for the 8 spaces, it will convert the number to an exponential format (%e).

The **display** command in its simplest form can be used to display the various formats. Using this simple command can be an ideal tool for exploring the possibilities of the various available formats.

```
display [%fmt] [=] exp
```

Other default display formats for the various data types include

| DATA TYPE | DEFAULT FORMAT |
|---|---|
| bytel | %8.0g |
| intl | %8.0g |
| longl | %12.0g |
| floatl | %9.0g |
| doublel | %10.0g |
| str#l | %#s |

The `%w.df` display format is a fixed format where *w* is the width of the display and *d* is the number of digits to the right of the decimal with the **f** indicating it is a fixed format. For example, **%6.2f** is a fixed format of six digits and two digits to the right of the decimal point.

### Experimentation Ideas

As I mentioned in the Introduction, this book deals primarily with agricultural research. Generally, this means using statistics to analyze planned experiments. There is a great body of statistical procedures that deal with exploratory analyses. This may be the beginning of a research project to tease out associations, differences, or trends. In addition, statistics are used to analyze large datasets of information for specific details that may be present. For example, census data may contain a great deal of information that is not immediately apparent. Statistics can help identify such information. In both of these cases, there is no formal experimental plan; however, there may be specific methods to acquire these data to ensure unbiased results. Much of agricultural research deals with planned experiments with carefully planned designs and treatment selection, which will be the emphasis of this book.

Biological systems by their nature will vary from one individual to another. This makes it difficult to determine whether treatment effects are real or just an artifact of these intrinsic differences. Even if two populations are treated exactly the same, they will differ when measured. For example, two plots of onions grown under the exact same conditions, when harvested will have different yields. These yields under these conditions obviously reflect no real difference. When an experiment is conducted and inevitably there are measured differences, are these differences real or do they reflect the intrinsic differences between individuals in a population? Various statistical procedures have been developed that give us a means of measuring and determining if these differences are real.

Data collected in experiments can be of a wide range of types. Numeric data can be parametric which means that it consists of a continuous range of numbers. An example would be the weight of experimental animals or the yield from vegetable plots. Other data may be nonparametric, which would include categorical data. These data would include things like sex (male or female). These two different types of data would use different statistical approaches for analysis. These different data types also are often referred to as continuous or discrete.

Ordinal data are yet another type of nonparametric or discrete data that use ranks. Ranked data would use yet another type of statistical approach for analysis. Data may include counts as well. Count data would be analyzed in a different method from continuous data.

In some cases, nonparametric or discrete data can be transformed to meet the criteria that would be used with parametric data (see Chapter 11). Or specific tests that were developed for nonparametric data can be used (see Chapter 12).

# 4
# Two Sample Tests

Simple statistical tests are available to determine if two means are different from one another. Such tests assume that the data are from a normal distribution, which, of course, is the famous bell-shaped curve. Two statistics can describe all such distributions, the mean and the variance.

One such statistic that can be used to determine if two means are different is the Z-test. This statistic does have some limitations and, in this context, it is rarely used. The primary limitation is the assumption that the population variance is known. In most cases, the entire population is not known. Instead, a sample from the population is used. This test can be used when sample sizes are large enough, which is seldom the case in planned experiments. Before the widespread use of computers, it used to be, as a rule of thumb, that sample sizes greater than 30 from a normally distributed population were sufficient to use the Z-test.

Stata does not supply the Z-test, per se, in the program, but it does calculate several density functions, one of which is the normal distribution of Z. Using the generalized formula below you can calculate a Z value and then compare it to the normal (Z) to see if it is significant.

$$Z = \frac{X - \mu_o}{\frac{\sigma}{\sqrt{n}}}$$

In this formula, X represents the mean of the value of interest. $\mu_0$ represents the population mean. The $\sigma$ value is the population standard deviation and the $n$ is the sample size. For example, using the Employee Salaries.dta, you can see how this works. This dataset consists of salaries for employees in the poultry industry. A random sample was obtained from a normally distributed population that consisted of salaries from poultry processing plants and feed mills. To begin with, let's use the **tabstat** command to display the means and standard deviations for this data. Enter

```
tabstat Salary, by(Source) stat(mean sd n)
```

which will display the following output:

```
Summary for variables: Salary
     by categories of: Source
     (1-processing plant, 2-feed mill)

  Source |      mean        sd         N
---------+------------------------------
       1 |  20.24749   8.48991        46
       2 |  17.21943  7.024119        35
---------+------------------------------
   Total |  18.93907  7.986932        81
----------------------------------------
```

At this point, we can use these output values to see if there is a real difference between feed mill employees and the entire poultry industry. The normal distribution will calculate a probability based on the Z value. To see this, enter in the Command window

```
display normal((17.22-18.94)/(7.99/sqrt(81)))
```

This results in a value of 0.026, which is below 0.05 (a commonly used critical probability value). This indicates that the feed mill employees on average are being paid less than employees overall. If the values 17.22 and 18.94 were reversed, the value would be 0.974, which gives the same value if subtracted from 1 (1 – 0.974 = 0.026). You might think that this is a long way to go to say there are differences in the salaries, which appears obvious. However, if you double the standard deviation from 7.99 to 16, the probability is 0.167, which is above 0.05, suggesting that there is no difference between the salaries.

Another use of Z values is determining an appropriate sample size in simple comparisons. If you look at the formula for the Z-test above, it is obvious it can be rearranged and solved for *n* (sample size). Stata does offer a command to calculate this.

```
sampsi #1 #2 [, options]
```

This command is capable of computing both sample size or power of the test. It is also capable of computing these values for one or two sample hypotheses as well as for both means and proportions. The *#1* and *#2* values are the proposed means. In addition, several options can be specified including the alpha level, power, sample sizes, and whether it is a one-sided or two-sided test, to name a few.

For example, a manufacturer of rolling greenhouse benches is thinking about changing its supplier of roller tubes. The new vendor says he can deliver 1 5/16-inch-diameter galvanized steel tube that is within 1/64 inch of this diameter. How large a sample would be needed to have a 95% confidence estimate with the mean diameter within these tolerances? Past data supplied by the vendor have the standard deviation at 1/32 inches. To answer this, enter

```
sampsi 1.3125 1.328125, sd1(0.03125) alpha(0.05)
power(0.5) onesample
```

The first number (1.3125) is the decimal form of 1 5/16 inches (pipe diameter) and the second number (1.328125) is 1 5/16 inches plus 1/64 inch in decimal form. The 0.03125 is the 1/32-inch standard deviation. The 95% confidence is entered as the alpha level (1 – 0.95 = 0.05) and the power (0.5) is the value entered to ignore the power. Many textbooks ignore the power when presenting this subject matter, which can be confusing. The results of this command are

```
Estimated sample size for one-sample comparison of
  mean to hypothesized value

Test Ho: m =  1.313, where m is the mean in the population

Assumptions:

         alpha =   0.0500  (two-sided)
         power =   0.5000
 alternative m =  1.32813
            sd =   .03125

Estimated required sample size:

             n =       16
```

The estimated sample size in this case is 16. If you were to enter a power of, say, 0.80, the result would be 32. In addition, this command can be used with two samples as well as with proportions.

This command also can be used to estimate the power of the test. It may be appropriate at this time to discuss some basic concepts in these types of statistics. Differences between means in a statistical context are determined by the probability of one mean occurring in the space of another. Statisticians often refer to this concept in the context of committing errors in favor of one mean over another. Two types of errors can be identified and are often referred to as type I and type II errors. A type I error, often denoted as α, is an error where the alternate hypothesis is wrongly chosen over the null or current hypothesis.

This may be better understood in the context of an experiment. Let's say a farmer is using a specific fertilizer rate and is producing his crop in a satisfactory manner (he's making money). In a statistical context, this fertilizer rate would be the null hypothesis or original mean. As a researcher, you think that a different rate may be better. This new rate would be considered the alternate hypothesis or new mean. Because your farmer is successful with what he is currently doing, as a researcher you don't want to recommend a different rate unless you are sure it will work. If you recommended a different rate and this was incorrect, that would be a type I error. See Figure 4.1 to see how this is represented graphically. As a researcher you want to minimize type I errors so the probability of committing this type of error is kept low. By convention, 5% or 1% are often used. In Figure 4.1, the type I error rate (or α) is shown as the area under the curve for the null

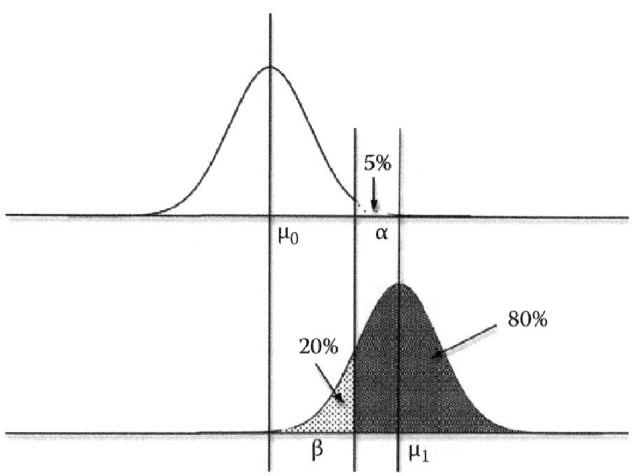


**Figure 4.1** Original mean ($\mu_0$) or null hypothesis compared to the new mean ($\mu_1$) or alternate hypothesis. α and β represent the type I and type II errors, respectively. 1-β, 80% in this case, represents the power of the test.

hypothesis or original mean ($\mu_o$) that represents 5%. There is a second type of error, the type II error that is often represented as β. A type II error is when the original mean ($\mu_o$) is selected when, in fact, the new mean ($\mu_1$) or alternate hypothesis is correct. We are not as concerned about this error because, as we said, the farmer is doing okay with his current fertilizer rate.

The power of a test is represented by 1-β, which is 80% in this case. This region of the new mean ($\mu_1$) represents the region that might be selected to find a difference or select the alternate hypothesis. If you slide the new mean ($\mu_1$) to the right, the power of the test increases and the ability to detect this new mean also increases. Conversely, if you slide this mean to the left, the power of the test is reduced.

Going back to our pipe example above and inputting a power value greater than 0.50 will change the results. A power value of 0.80 results in a sample size of 32 and a power of 0.90 requires 43 samples.

As mentioned earlier the Z-test has very limited usefulness because it requires that the population variance be known and this is rarely the case. For two sample means, the t-test is more often used. Stata offers several methods of computing a t-test for both one-sample and two-sample datasets with either paired or unpaired data. In addition, Stata has an immediate form of the t-test that does not require a dataset for computation.

For a one-sample t-test, enter the following command:

```
ttest varname == # [if] [in] [, level(#)]
```

The `varname` is the variable in your dataset you wish to analyze and the # is the arbitrary mean to compare with. The **if**, **in**, and **level(#)** are optional. **if** and **in** allow a selection of the observations to be used while **level(#)** can be used to set the confidence level, which by default is set to 95.

In onion production, an average yield is about 500 40-lb boxes per acre. This translates into about 55 lbs/plot (120 ft$^2$). One way this command could be used would be to compare this average yield to the actual yield from an experiment. Open the onionyield03.dta dataset and enter the following command:

```
ttest fieldyield == 55
```

This results in the following output:

```
One-sample t test
------------------------------------------------------------------------------
Variable |     Obs        Mean    Std. Err.   Std. Dev.   [95% Conf.Interval]
---------+--------------------------------------------------------------------
fieldy~d |     120    93.82417    2.468546    27.04157     88.9362   98.71213
------------------------------------------------------------------------------
    mean = mean(fieldyield)                                       t =  15.7275
Ho: mean = 55                                    degrees of freedom =      119

    Ha: mean < 55               Ha: mean != 55                Ha: mean > 55
 Pr(T < t) = 1.0000         Pr(|T| > |t|) = 0.0000         Pr(T > t) = 0.0000
```

From this output, we can see that the mean from this experiment (93.8 lbs/plot) is considerably higher than the average yield of 55 lbs/plot. The t is the calculated t value, which is used to determine a probability of statistical significance. At the bottom of this output table are three listed probabilities (Pr). The first indicates there is no chance the mean is lower than 55. The second indicates that there is a significant difference between the calculated mean and 55, while the last indicates the calculated value is significantly greater than 55. This may seem unimportant in this case, but there are cases where the researcher is specifically interested in whether a value is above or below a specified value.

The two-group t-test determines if there are differences between two groups. Open the file simplepumpkin.dta and enter the command

```
ttest yield, by (variety)
```

This command compares two means using variety as a grouping variable; in this case, two different pumpkin varieties. The yield variable represents the yield from four plots from each variety.

The output is slightly different than displayed with the previous use of this command, but will give information similar to the above output.

```
Paired t test
------------------------------------------------------------------------------
Variable |     Obs        Mean    Std. Err.   Std. Dev.   [95% Conf. Interval]
---------+--------------------------------------------------------------------
  normal |       4     286.225    41.15072    82.30145     155.265     417.185
orange~l |       4     356.525    132.1471    264.2942    -64.0261    777.0761
---------+--------------------------------------------------------------------
    diff |       4       -70.3      159.65       319.3   -578.3775    437.7775
------------------------------------------------------------------------------
     mean (diff) = mean(normal - orangebull)                      t =  -0.4403
 Ho: mean (diff) = 0                             degrees of freedom =        3

 Ha: mean (diff) < 0         Ha: mean(diff) != 0           Ha: mean(diff) > 0
 Pr(T < t) = 0.3447        Pr(|T| > |t|) = 0.6895          Pr(T > t) = 0.6553
```

In this case, the difference between the two varieties, PMK-06-04 and Orange Bulldog, is not significant. The output lists three different probabilities, T<t, |T|>|t|, and T>t. The first and last values, 0.3447 and 0.6553, are one-tailed probability values for the t-test, while the middle value, 0.6895, is a two-tailed probability value. Note that the first and third probability values add to 1.

There are two other forms of the t-test available: the paired and unpaired. To see these forms, open the file ttestpumpkin.dta and look at the data. It is the same data as in simplepumpkin.dta, but the data have been entered in two columns. If the command is entered `ttest normal == new`, it is assumed to be paired and, although in this case the results would be the same, it is technically incorrect because it assumes the two columns are paired and, in fact, they represent completely different plots. The command should be entered as `ttest normal == new, unpaired` to be correct. Paired data would be used where treatments are applied to the same experimental units or you are interested in the difference between the paired values. For example, in a study examining weight gain in steers with and without a new wormer, weight gain may be recorded before administering the new drug and again after administration. Here, there is a clear association between each pair of measurements because both are observed in each animal.

Finally, there is an immediate form of the t-test available for both one-sample and two-sample cases.

An example for a one-sample case would be to enter the values

```
ttesti 120 93.8 27.0 55
```

The first value is the number of observations (120), next is the test mean (93.8), the third value is the standard deviation (27.0), and, finally, the value (55) is the test mean with which to compare.

## ANOVA

One of the most important types of analysis used is the analysis of variance (ANOVA). This expands beyond the t-test by offering a method to analyze more than two sample means. In this type of analysis, a specific value, called F, is calculated. This is named in honor of R. A.

Fisher (English statistician, 1890–1962) who first proposed this type of testing. There is a relationship between this analysis and the t-test where only two treatments are involved. If the t-test value is squared, it will equal the F value from an ANOVA. In this type of analysis (in its simplest form), variances are calculated for experimental units treated in the same fashion, often called the *within group variance* and for treatments treated differently called the *between group variance*. If the treatments are different, then the between group variance will be larger than the within group variance. This usually is represented as a ratio of the between group variance over the within group variance and is called *F*. A probability is then calculated to indicate what is the chance of this difference occurring by chance alone. By custom, probabilities of 0.01, 0.05, or 0.10 are often used to declare treatment effects having a statistical difference.

This simple experimental design is called a completely randomized design (CRD). With this design, treatments are assigned randomly to experimental units. This type of experimental design is used where there is a great deal of uniformity between the experimental units other than treatment effects or there is very little difference between experimental units because of environment or location. Examples where this type of design might be used include a growth chamber where conditions other than the treatments would be very uniform. Greenhouse experiments also may be arranged in this fashion, although there are often sufficient differences between locations in a greenhouse to warrant the use of a different experimental design. Finally, in animal experiments where the animals (experimental units) are considered reasonably uniform before treatment application can be tested with a completely randomized design. Animal uniformity might include selecting animals with similar weights or ages and, in addition, no attempt is made to segregate the animals.

There are several ANOVA commands within Stata that can be used to analyze this type of design. They include **oneway**, **loneway**, and **anova**. Each arrives at the same solution, while each offers slightly different information. At this time, we will concentrate on **oneway** and **loneway** because they restrict the model to just one independent variable. The oneway command can be entered as follows:

```
oneway response_var factor_var [if] [in] [weight] [, options]
```

The `response _ var` is the dependent variable, which is what is being measured, while the `factor _ var` is the independent variable or the specific treatments. There are several options available with this command including the ability to produce several multiple-comparison tests.

### Output and Meaning

Open the dataset virustest.dta. This is a sample dataset of watermelon plants that were inoculated with zucchini yellow mosaic virus—Egyptian strain (ZYMV-E). This virus is particularly virulent on watermelons and was not found in the United States. Because of this, to prevent this virus from escaping during testing, evaluation was restricted to a growth chamber. Because growth chambers have very uniform conditions, the experiment was set up as a CRD of 11 plants (replications) of each of nine watermelon cultivars.

Enter the following command:

```
oneway absorb trt
```

This will result in the following:

```
                        Analysis of Variance
    Source          SS          df        MS             F      Prob > F
------------------------------------------------------------------------
Between groups  3.12941196   8   .391176495     145.97      0.0000
 Within groups   .241184086  90   .002679823
------------------------------------------------------------------------
    Total       3.37059605   98   .034393837

Bartlett's test for equal variances:
  chi2(8) = 21.5128 Prob>chi2 = 0.006
```

The resulting table is called an *analysis of variance* table. The presentation of ANOVA calculations in this form is fairly standard across textbooks and statistical programs. The columns, Source, SS', df', MS', F, and Prob > F will be present in all ANOVA tables, but the number of rows will differ as the experimental designs become more complex resulting in more complex models. This will become evident as we look at more complex designs later in the book.

In the CRD under Source is listed Between groups, which represents the treatments or, in this case, the watermelon varieties. The Within groups represents the experimental error or differences that occur due to such things as minor errors in measurement or the natural differences that occur between individuals.

The next column, SS, is the abbreviation for *sum of squares*, which is followed by the df column. The df stands for *degrees of freedom* and represents one less than the number of items in this source of variation. In this case (Between groups), there were nine cultivars, thus, the number listed is (9 – 1) or 8. The Within groups degrees of freedom is 90, which is the total of all the degrees of freedom for each cultivar. The total number of experimental units in this study was 99, so the total degrees of freedom is 98.

The next column, MS, is the *mean square* column and this is calculated by dividing the sum of squares by the degrees of freedom. The mean squares listed are variances, from which is calculated the F value. The F value is the Between groups variance or mean square divided by the Within groups mean square. The *Prop > F* is the probability of the F value occurring by chance alone. In this case, a value of 0.0000 indicates that there is a real difference based on a 0.01 or even a 0.001 threshold.

The last line in the table calculates Bartlett's test for equal variances. One of the underlying assumptions with ANOVA is that the variances between the treatments (cultivars) be the same, and, in this case, they are not. For the time being, we will ignore this and come back to it in a later chapter.

As mentioned earlier, there is more than one command that can calculate an ANOVA. The **`loneway`** command can do the same calculation. The **`loneway`** command is primarily used for large one-way ANOVAs. The **`loneway`** command can be used to calculate ANOVAs with levels (treatments) greater than 376, while the **`oneway`** can only calculate experiments up to 376 levels.

Below is the output from the **`loneway`** command of the same virus screening data. Note the ANOVA table is the same, but there are again more data present. First there is the addition of an R-squared value. This is a value from 0–1 that reflects how well the treatments predict the outcome and, in this case, is calculated as the between treatment sum of squares divided by the total sum of squares. The closer this value is to 1, the better the model fits.

The other values presented are rarely if ever presented in agricultural experiments; however, a brief explanation is in order. The Intraclass correlation is the upper bounds of the `response _ var` explained by the `group _ var` and is a similar evaluation of the data to the R-squared. The asymptotic standard error (Asy. S.E.) and 95% confidence interval relate to the reliability or dispersion of the intraclass correlation. The estimated standard deviation (SD) of between and within treatment effects can be compared to the square root of the mean square values in the ANOVA table. In the case of the estimated SD between treatments, a percent of the treatment effect can be seen by comparing it to the square root of the mean square for the between treatment effect in the ANOVA table.

```
One-way Analysis of Variance for absorb:

                                    Number of obs =          99
                                        R-squared =      0.9284

    Source            SS          df        MS          F       Prob > F
-------------------------------------------------------------------------
Between trt       3.129412        8     .3911765    145.97    0.0000
Within trt       .24118409       90     .00267982
-------------------------------------------------------------------------
Total             3.370596       98    .03439384

         Intraclass        Asy.
         correlation       S.E.          [95% Conf. Interval]
         -----------------------------------------------------
            0.92947      0.03444         0.86198      0.99697

         Estimated SD of trt effect                  .1879305
         Estimated SD within trt                      .051767
         Est. reliability of a trt mean               0.99315
              (evaluated at n=11.00)
```

# 5

# Variations of One Factor ANOVA Designs

## Randomized Complete Block Design

In the previous chapter, ANOVA (analysis of variance) was introduced with the simplest of experimental designs, the completely randomized design, which is analyzed with the one-way ANOVA. However, there can be more than one predictive or treatment effect in a design. Probably the most common method of analyzing agricultural experiments is the randomized complete block design (RCBD). In this design, replications are arranged into blocks to reduce experimental error that may occur because of differences in field location. Often the terms *replication* and *blocks* are used interchangeably so that a researcher might refer to an experiment with x treatments and y replications with one replication occurring in each block. There are experiments, as we will see later, that can have both replications and blocks. Keep this in mind so you won't be confused as we look later at more complex designs.

The command we will be using to look at these evermore complex designs is **anova**, which has the form

```
anova varname [term [/] [term [/]...]][if] [in]
[weight] [, options]
where term is of the form varname[{#||}varname[...]]
```

The **anova** command is followed by *varname*, which is the dataset dependent variable. This is what is actually measured. The *term*, of which there can be several, is the independent variable or the treatment effects. Each *term* can have several variables (*varname*) and interactions and/or subcategories. Interactions are noted with # and subcategories with |.

Many commands, including **anova,** have the general form as listed above and can include modifiers, such as ***if***, ***in***, and ***weight***, that further restrict or define how the independent variables will be used in the analysis. ***if*** conducts the analysis on only the data as restricted by this modifier. The `in` modifier allows you to restrict variables by a range of observations. Finally, **weight** can be specified if the data are weighted in some fashion. With **anova** there can be a frequency or analytic weight. In the former, the weighting indicates the number of duplicate observations, while the latter are weights indicating the inverse proportion of the weight to the variance of the form

$$\frac{\sigma^2}{w_j}$$

where $\sigma^2$ is the variance of an observation and $w_j$ is the weight for the $j$th observation.

Load the dataset Onion trial 1999.dta into Stata and enter the following command:

```
anova yieldacre entry rep
```

This will result in the following output:

```
Number of obs =        60     R-squared     =  0.7882
Root MSE      = 101.464       Adj R-squared =  0.6712

  Source |  Partial SS    df        MS            F      Prob > F
 --------+---------------------------------------------------------
   Model |  1455957.67    21   69331.3176       6.73      0.0000
         |
   entry |  1433977.53    19   75472.5015       7.33      0.0000
     rep |  21980.1412     2   10990.0706       1.07      0.3539
         |
 Residual|  391207.018    38   10294.9215
 --------+---------------------------------------------------------
   Total |  1847164.69    59   31307.8761
```

This dataset is an onion variety trial that had 20 different varieties evaluated with three replications arranged in an RCBD. The yieldacre is the extrapolate yield/acre in 50-lb bags per acre, which was calculated from the yield variable. The individual plot size or experimental unit was 120 $ft^2$ and the yield per plot is listed in the yield variable. I often extrapolate such data into the units that will be most useful in either publications or for grower meetings. Using 50-lb bags/acre is a common method for presenting onion data. Depending on the situation and crop, you may wish to calculate boxes/acre, lbs/acre, or some other common unit. It makes it easier when calculating and presenting results in tables and graphs. Whether the analysis is conducted on the raw results (lbs/plot) or the extrapolated results, this will not change the outcome.

There are several pieces of information presented in the ANOVA table. The number of observations (60) reflects the total number of experimental units, 20 varieties with three replications each. The R-square ($R^2$) is the same value discussed previously with the one-way ANOVA and, in this case, it is the Model sum of squares divided by the Total sum of squares. This is often referred to as the coefficient of determination with the following formula:

$$R^2 = 1 - \frac{SS_{\text{Residual}}}{SS_{\text{Total}}}$$

The $R^2$ will increase in value as more independent variables are added, but may not truly reflect an increase in its predictive ability. To compensate for the phenomenon, an adjustment to the $R^2$ value has been proposed with the formula

$$R_a^2 = 1 - \frac{MS_{\text{Residual}}}{MS_{\text{Total}}}$$

This value will always be lower than the $R^2$ and compensates for the number of independent factors in the model, thus, with models with several independent factors, this may better reflect the actual predictive nature of the model.

The Root MSE is the square root of the mean square error or the residual mean square. In this case, it is $\sqrt{10294.9215}$, which is 101.464. The remainder of the table is much as it was described in the one-way ANOVA described previously. There are, however, more rows listed in the RCBD. The Model is an estimate of the combined entry and rep sources in the experiment. The entry in this case are the varieties, the rep are the blocks with one replication of each variety in each rep, and the Residual is the error or background noise that occurs in the experiment. The Residual is important in this design because it is the denominator in the calculated F-tests. The only really important F-test is for entry where we see a highly significant difference between varieties with a Prob > F of 0.0000.

Because the rep source of variation was not significant, there is not much difference between calculating this model as an RCBD or a CRD (completely randomized design). This is not always the case; the blocking effect (rep), when significant, can account for a lot of variation in the model. This accounted-for variation can lower the Residual mean square making it more likely to detect differences between the treatments. In fact, it is possible to calculate the relative efficiency of RCBD compared to a CRD by the formula

$$R.E. = \frac{(r-1)E_b + r(t-1)E_e}{(rt-1)E_e}$$

In this formula, the $r$ represents the number of replications, which is 3 in this case. $E_b$ is the replication mean square, which is 10990.0706, and $E_e$ is the residual mean square, which is 10294.9215. The $t$ is the number of treatments, which in this case is 20.

In Chapter 3, it was mentioned that some commands save results for further calculations. The **summarize** command was used as an example saving several results in `r()`. The **anova** command also saves results, but these results are saved in `e()`, which is used by e-class commands, estimation commands. Type in **ereturn list**, which should be entered immediately after the **anova** command:

```
ereturn list
```

The following results will appear:

```
scalars:
                    e(N)  =  60
                 e(df_m)  =  21
                 e(df_r)  =  38
                    e(F)  =  6.734516384698626
                   e(r2)  =  .7882121609461262
                 e(rmse)  =  101.4638926803575
                  e(mss)  =  1455957.670464247
```

```
                   e(rss) =  391207.0176783416
                  e(r2_a) =  .6711715130479328
                    e(ll) =  -348.6157398005913
                  e(ll_0) =  -395.1808477381046
                  e(ss_1) =  1433977.529230501
                  e(df_1) =  19
                   e(F_1) =  7.331041951857553
                  e(ss_2) =  21980.14123374692
                  e(df_2) =  2
                   e(F_2) =  1.067523496688828

macros:
              e(cmdline) : "anova yieldacre entry rep"
               e(depvar) : "yieldacre"
                  e(cmd) : "anova"
           e(properties) : "b_nonames V_nonames"
             e(varnames) : "entry rep"
               e(term_2) : "rep"
               e(term_1) : "entry"
               e(sstype) : "partial"
              e(predict) : "regres_p"
                e(model) : "ols"
            e(estat_cmd) : "anova_estat"

matrices:
                     e(b) :  1 x 24
                     e(V) :  24 x 24

functions:
              e(sample)
```

This information can be used to calculate the relative efficiency of the RCBD compared to the CRD. We can use the **display** command to calculate the relative efficiency.

Enter the following commands:

```
local Eb = e(ss_2)/e(df_2)
local r = e(df_2)+1
local t = e(df_1)+1
local Ee = e(rss)/e(df_r)
local RE = ((`r'-1)*`Eb' + `r'*(`t'-1)*`Ee')/
((`r'*`t'-1)*`Ee')
display `RE'
```

Each of the first five lines are calculated local variables or, in the vernacular of Stata, local macros. The last line displays the results of the last calculated macro (RE). If you are familiar with other programming languages, Stata's use of macro is somewhat different. The *local* term is required to distinguish these macros (variables) from global macros. The values assigned to these macros are calculated from the list of scalars (numbers) from the previously executed estimation command (**anova**) and listed with the **ereturn list** command. You will notice in the last two lines that these local macros have `and ' quotes around them. This is different from other programming languages as it indicates to the Stata program to use the value of the macro rather than the macro name. With the ` and ' around `r', it recognizes the value (3); without the ` and ' it would recognize it as r. The ` is an accent mark located at the upper left side of most keyboards. The other (') is a closed quote found near the return key. We will look more closely at this in Chapter 7 on programming.

It is important to remember that the values listed in the **ereturn list** are only available until the next estimation command is executed. If you were to calculate another ANOVA, the values would change to the new estimation.

In this case, the R.E. is 1.0022889, which is quite small. It means by using an RCBD instead of the CRD we are only seeing 0.2% increase in efficiency. This will not always be the case; in fact, the relative efficiency can be quite large in some cases. Finally, if the error degrees of freedom or Residual degrees of freedom is below 20, a correction factor should be calculated and multiplied against the relative efficiency. In this case, with an error degrees of freedom of 38, it is not necessary; however, for those cases where it would be required the correction factor is

$$k = \frac{\left[(r-1)(t-1)\right]\left[t(r-1)+3\right]}{\left[(r-1)(t-1)+3\right]\left[t(r-1)+1\right]}$$

### Latin Square Designs

Latin square (LS) designs add another source of variation and, hence, consist of both row and column variations. This design requires that the number of treatments equal the number of rows and columns in the design. Often this design is used where two different gradients may occur in a field, perhaps soil fertility in one direction and soil moisture in another. The limitation that the number of treatments equals the number of columns and rows limits the usefulness of such designs because as the number of treatments increases, the number of experimental units can quickly increase to an unwieldy number. Often with a small number of treatments, four or less, two identical LSs are used to increase the precision of the experiment. Generally LSs with more than eight treatments are not conducted because of the unwieldy nature and size of such experiments.

Load the data file Latin square 1.dta into Stata and enter the following command:

```
anova pyr trt row column
```

This dataset represents an experiment where the variable trt represents four different rates of sulfur fertilizer (0, 20, 40, 60 lbs/acre) applied to onions as part of a complete fertilizer program. The row and column variables represent the plot position in the experiment. The pyr variable represents the pyruvate value measured from 10 bulbs from each experimental unit. The pyruvate test is a relative measure of onion pungency. This results in the following output:

```
Number of obs =        16     R-squared     =  0.9446
Root MSE      =  .260192     Adj R-squared =  0.8615

  Source |  Partial SS    df       MS            F     Prob > F
 --------+------------------------------------------------------
   Model |  6.92497434     9  .769441593       11.37     0.0039
         |
     trt |  2.96342539     3  .987808464       14.59     0.0037
     row |   .38727494     3  .129091647        1.91     0.2296
  column |    3.574274     3  1.19142467       17.60     0.0022
         |
Residual |  .406200239     6   .06770004
---------+-------------------------------------------------------
   Total |  7.33117457    15  .488744972
```

Unlike the previous analysis, there are now two additional sources of variation; in addition to the treatment effect (trt), there are row and column effects. In the previous experiment, the rep variable represents blocks in the field that are used to account for potential differences that may occur because of location. In this particular case, the LS accounts for two additional sources of variation, both row and column position. From a research perspective, the fertilizer treatments do have an effect on bulb pyruvate with a significance of Prob > F of 0.0037, which would be of primary interest.

The column variable is significant, but the row variable is not. It is possible to calculate the relative efficiency of the LS design compared to a CRD as well as the RCBD. These formulas include

$$R.E.(CRD) = \frac{E_r + E_c + (t-1)E_e}{(t+1)E_e}$$

$$R.E.(RCB, row) = \frac{E_r + (t-1)E_e}{(t)(E_e)}$$

$$R.E.(RCT, column) = \frac{E_c + (t-1)E_e}{(t)(E_e)}$$

There is also a correction factor for the LS design that should be used if the error degrees of freedom is below 20. In this case it is at 6. The correction factor is

$$k = \frac{\left[(t-1)(t-2)+1\right]\left[(t-1)^2+3\right]}{\left[(t-1)(t-2)+3\right]\left[(t-1)^2+1\right]}$$

The following listing uses the scalars from the previous estimation and calculates the relative efficiencies of the LS design compared to CRD and RCBD designs. Entering this sequence of commands can be tedious and error prone, so I have already done it for you. Open the Do-File LS Efficiency.do in the Do-File folder. Once open in a Do-file Editor, click the Do icon in the upper right-hand corner of the editor. This will display the four calculated values: k, RE, RER, and REC, which represent the correction factor, relative efficiency, relative efficiency of the rows, and relative efficiency of the columns, respectively. The code segment is an example of a Do-File that will be discussed in more detail in Chapter 7. Suffice it to say, this is one of the great strengths of Stata—its extensibility.

```
local Er = e(ss_2)/e(df_2)/*Row mean square*/
local Ec = e(ss_3)/e(df_3)/*Column mean square*/
local t = e(df_1)+1/*Number of treatments*/
local Ee = e(rss)/e(df_r)/*Error mean square*/
local k = ([(`t'-1)*(`t'-2)+1]*[(`t'-1)^2+3])/([(`t'-1)*
(`t'-2)+3]*[(`t'-1)^2+1])/*Correction factor*/
local RE = (`Er' + `Ec' + (`t'-1)*`Ee')/((`t' + 1)*`Ee')*`k'
local RER = (`Er' + (`t'-1)*`Ee')/(`t' * `Ee')*`k'
local REC = (`Ec' + (`t'-1)*`Ee')/(`t' * `Ee')*`k'
display `k'
display `RE'
display `RER'
display `REC'
```

Remember, as before, the scalars are only available from the most recently executed estimation command. The results of executing this Do-File are a correction factor of 0.93 and relative efficiencies of 4.20, 1.14, and 4.81 comparing the LS to the CRD, RCBD (rows), and RCBD (columns), respectively. This means that the LS design is 320% more efficient than the CRD or the CRD would require 3.2 times more replications to attain the efficiency of the LS design. In addition, the LS design has increased precision with the row blocking of 14% and column blocking of 381%.

With small LS designs two identical experiments can be conducted and the results analyzed together. Load the dataset Latin square 2.dta for the next analysis. This dataset represents an experiment with three different initial fertilizer applications as part of an overall fertilization program with direct seeded onions. The initial fertilizer treatments were 0 fertilizer, 150 lbs/acre calcium nitrate ($Ca(NO_3)_2$), and 200 lbs/acre diammonium phosphate ($(NH_4)_2H_2PO_4$). Each treatment was applied to a single row, so, in addition to the replication source of variation, there were individual fertilizer hoppers.

Enter the following command to analyze these data. Note the vertical bars in this command. This key (|) is usually found just above the return key on most keyboards:

```
anova yield exp rep|exp hop|exp trt
```

**Table 5.1** Source of variation and degrees of freedom for two Latin square experiments conducted simultaneously

| SOURCE OF VARIATION | DEGREES OF FREEDOM (DF) | RESULTING DF |
|---|---|---|
| Experiments (*e*) | $e - 1$ | $2 - 1 = 1$ |
| Replications within experiments | $e(r - 1)$ | $2(3 - 1) = 4$ |
| Hoppers within experiments | $e(h - 1)$ | $2(3 - 1) = 4$ |
| Treatments (*t*) | $t - 1$ | $3 - 1 = 2$ |
| Error | $(et - e - 1)(t - 1)$ | $((2)(3) - 2 - 1)(3 - 1) = 6$ |
| Total | $et^2 - 1$ | $2(3)2 - 1 = 17$ |

This will result in the following output:

```
Number of obs =        18     R-squared     =  0.8246
Root MSE      =    2.4226     Adj R-squared =  0.5030

  Source |  Partial SS    df       MS             F     Prob > F
  -------+-------------------------------------------------------
   Model |  165.543835    11  15.0494395        2.56     0.1294
         |
     exp|  15.5558083     1  15.5558083        2.65     0.1546
 rep|exp|  114.810878     4  28.7027195        4.89     0.0426
 hop|exp|   24.581224     4    6.145306        1.05     0.4561
     trt|  10.5959244     2  5.29796218        0.90     0.4542
         |
Residual |  35.2139522     6  5.86899204
---------+-------------------------------------------------------
   Total |  200.757787    17  11.8092816
```

The vertical bars (|) indicate the variable preceding it is nested in the variable just after the bar. This is best illustrated with a table showing the sources of the degrees of freedom (Table 5.1).

Wherever you see a source of variation nested within another variable, such as [$e(r - 1)$], reverse their position and place the bar character between as indicated above (rep|exp). In this experiment, there were no treatment effects with an F value of 0.90 and Prob>F of 0.4542. The replication within an experiment was significant, while the hopper within an experiment was not and we would expect the relative efficiency to have been increased by replications, but not by the hopper as a source of variation.

## Balanced Incomplete Block Designs

Balanced incomplete block designs are marked by having less experimental units than there are treatments to be tested within a block.

These designs are considered balanced because pairs of treatments occur an equal number of times in the experiment. In addition, each treatment is replicated the same number of times throughout the experiment. Incomplete block designs are usually employed because of some limitation in space or equipment to test all the treatments within each block.

Balanced incomplete block (BIB) designs will have $k$ experimental units, $b$ blocks, $t$ treatments, $r$ replications, and $\lambda$ blocks in which pairs of treatments occur. The total number of experimental units, $N$, $= rt = bk$. In addition, these designs will have $k < t$. Finally, $\lambda = r(k-1)/(t-1)$ in these experiments.

Load the dataset Broccoli Germination.dta. This is a dataset of broccoli root growth as a function of temperature. There were only three growth chambers ($k = 3$) available to control the temperature and the experiment called for four germination temperatures ($t = 4$). The experiment was set up as a BIB design with four temperature treatments of 10, 20, 30, and 40°C. Root growth was measured as $mm \cdot h^{-1}$. Four separate runs ($b = 4$) were conducted in such a fashion so that each temperature occurred three times ($r = 3$) in the experiment and was paired with each other temperature within a run twice ($\lambda = 2$).

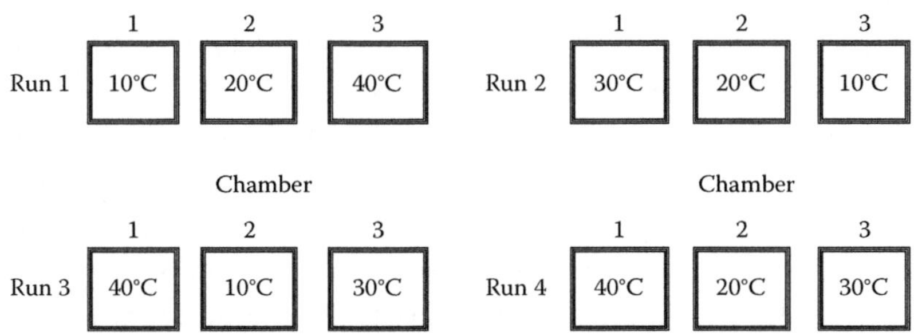


If you look at the dataset in the Data Editor window you will notice that there are missing data indicating the design is incomplete. It is, however, balanced because each treatment occurs the same number of times and is paired with other temperatures the same number of times. To meet these requirements, particularly as the number of treatments increase, can become quite complex and it is best to consult a textbook or statistician before proceeding.

Enter the following command and look at the resulting output:

```
anova root run temp, sequential

          Number of obs = 12          R-squared     =  0.9544
          Root MSE      = .074633   Adj R-squared =  0.8996

   Source |     Seq. SS      df          MS             F     Prob > F
  --------+----------------------------------------------------------
    Model |  .582504477       6    .09708408        17.43       0.0033
          |
      run |  .115075577       3   .038358526         6.89       0.0317
     temp |  .467428901       3   .155809634        27.97       0.0015
          |
 Residual |  .027850417       5   .005570083
 ---------+----------------------------------------------------------
    Total |  .610354894      11   .055486809
```

It is important to enter the command exactly as listed above including the order of the variables. A sequential sum of squares is often referred to as a type I sum of squares and the partial sum of squares as the type III sum of squares or sometimes as the adjusted sum of squares. Because not all observations occur simultaneously (we have only three growth chambers and four treatments), the order of the calculations is important. The sum of squares accounted for by temperature is calculated after taking into account the sum of squares for the runs. Normally the **anova** command defaults to the partial sums of squares where the order of the independent variables does not matter. Since the **anova** command defaults to the partial sums of squares, it does not have to be explicitly listed as an option. The sequential sum of squares calculated the first sum of squares, which influences the subsequent calculation. Try it for yourself by reversing the order of the run and temp variables. In addition, whether the `sequential` option is specified or not will change the results. From this ANOVA the germination temperature is significant, but the run is significant as well. Because of this, the least squares means should be reported rather than the arithmetic means when reporting these results. To calculate the least squares or marginal means enter the following command:

```
margins temp
```

This will result in the following output:

```
Predictive margins                          Number of obs     =         12

Expression    : Linear prediction, predict()

--------------------------------------------------------------------------
      |            Delta-method
      |     Margin   Std. Err.      z    P>|z|     [95% Conf. Interval]
------+-------------------------------------------------------------------
 temp |
  10  |   .0235833   .0450639     0.52   0.601   -.0647404     .111907
  20  |   .4334583   .0450639     9.62   0.000    .3451346    .5217821
  30  |   .4928333   .0450639    10.94   0.000    .4045096     .581157
  40  |   .0724583   .0450639     1.61   0.108   -.0158654    .1607821
--------------------------------------------------------------------------
```

The first column labeled Margin contains the marginal or least squares means. To compare it to the arithmetic means, enter

```
mean root, over(temp)
```

which results in

```
Mean estimation                              Number of obs     =        12

           10: temp = 10
           20: temp = 20
           30: temp = 30
           40: temp = 40

--------------------------------------------------------------
        Over |       Mean   Std. Err.     [95% Conf. Interval]
-------------+------------------------------------------------
root         |
          10 |   .0023333    .001453     -.0008646    .0055313
          20 |   .4366667    .031798      .3666798    .5066535
          30 |   .5033333   .0768837      .3341133    .6725533
          40 |        .08   .0152753      .0463794    .1136206
--------------------------------------------------------------
```

Although there is not a great deal of difference between the two sets of means, there are differences particularly for the 10°C treatment.

Finally, if you open the Data Editor window and look at this dataset, you will notice missing values for dependent variable, root. These observations do not have to be part of the dataset for the calculations to work correctly. They were left in to emphasize the fact that the design is incomplete. The chamber variable also is not required for the calculations and just indicates that there were only three growth chambers available.

With previously discussed models, it was possible to calculate a relative efficiency of a more complex design to a simpler design. Such comparisons with BIB designs are not directly possible because the designs are incomplete. If the number of treatments and replications is the same, however, between the BIB and RCBD, then the ratio of the variances of the difference between two treatment means for the RCBD and BIB is an indication of efficiency.

$$Efficiency = \frac{\left(2\sigma^2_{rcb}/r\right)}{\left(2k\sigma^2_{bib}/\lambda t\right)} = \frac{\sigma^2_{rcb}}{\sigma^2_{bib}} \cdot \frac{\lambda t}{rk}$$

Assuming the variances between the RCBD and BIB designs are the same, then

$$E = \frac{\lambda t}{rk}$$

and indicates the loss in efficiency by using the BIB design relative to the RCBD. For example, an experiment with six treatments and five replications and four experimental units would have $t = 6$, $r = 5$ for the RCBD and $t = 6$, $r = 5$, $k = 4$, $\lambda = 3$ for the BIB, which would result in E = 0.9. This means the amount the BIB variance would have to be reduced relative to the RCBD design would be about 10% for the same efficiency.

### Balanced Lattice Designs

As the number of treatments increases, there is a concomitant increase in the size of blocks. This can lead to blocks that are not very uniform for treatment conditions. From one side of a field to the other, as these distances increase, the chance for conditions in soil type, moisture, fertility, etc., to change increases. Balanced lattice designs address this problem with blocks of relatively small size and increasing the number of replications as the number of treatments increases. This can be confusing if you are used to using RCBDs. In RCBDs, the terms *blocks* and *replications* are synonymous and are used interchangeably. In the context of balanced lattice designs, they are distinct sources of variability.

Balanced lattice designs have several required constraints. First, the number of treatments must be a perfect square, such as 9, 16, 25, 36, etc. This at first seems like a pretty strong restriction on the experimental design, but it is usually easy to include a couple of extra treatments or to delete some. These designs were for relatively large experiments, thus, in this context this is not much of a restriction. One place where these types of experiments are used is in testing large numbers of potential new varieties. Plant breeders may be interested in looking at many advanced lines, and balanced lattice designs are a good choice for this.

Using $k$ as the root value for these designs, the number of treatments is the square of $k$, ($t = k^2$), the number of replications must be $r = (k + 1)$, and the number of blocks are $b = k(k + 1)$ with $\lambda = 1$ being the number of treatment pairs within a block. Consult a good statistical text, statistician, or the Internet for layouts of these designs.

Open the file Lattice design.dta, which is in a separate folder within the Data folder, Lattice. This is a file with 16 fertilizer treatments and their effect on tiller number in rice (Gomez and Gomez, 1984, p. 45). There are two other sources of variability other than the treatments within this file, which include blocks (block) and replications (rep). Before working through this example, it is a good idea to change the working directory to the Lattice folder. This can be done from the File menu by choosing Change Working Directory… and selecting this folder or you can use the `cd` command and enter the path to this folder.

The layout of the experiment is shown in Table 5.2. All 16 treatments are present in each replication and each row within the replications is blocks. Also, each treatment pair occurs once within a block. For example, treatment 1 occurs with treatment 3 (paired) in block 1, but these treatments do not occur together in a block anywhere else in the experiment. Table 5.3 shows the degrees of freedom and which mean squares are used to calculate the F values.

**Table 5.2** Layout of lattice design experiment for fertilizer treatment effect on rice tiller number

| BLOCK NUMBER | | | | REPLICATION I | | | | |
|---|---|---|---|---|---|---|---|---|
| 1 | *1** | **147†** | *2* | **152** | *3* | **167** | *4* | **150** |
| 2 | *5* | **127** | *6* | **155** | *7* | **162** | *8* | **172** |
| 3 | *9* | **147** | *10* | **100** | *11* | **192** | *12* | **177** |
| 4 | *13* | **155** | *14* | **195** | *15* | **192** | *16* | **205** |
| | | | | REPLICATION II | | | | |
| 5 | *1* | **140** | *5* | **165** | *9* | **182** | *13* | **152** |
| 6 | *10* | **97** | *2* | **155** | *14* | **192** | *6* | **142** |
| 7 | *7* | **155** | *15* | **182** | *3* | **192** | *11* | **192** |
| 8 | *16* | **182** | *8* | **207** | *12* | **232** | *4* | **162** |
| | | | | REPLICATION III | | | | |
| 9 | *1* | **155** | *6* | **162** | *11* | **177** | *16* | **152** |
| 10 | *5* | **182** | *2* | **130** | *15* | **177** | *12* | **165** |
| 11 | *9* | **137** | *14* | **185** | *3* | **152** | *8* | **152** |
| 12 | *13* | **185** | *10* | **122** | *7* | **182** | *4* | **192** |
| | | | | REPLICATION IV | | | | |
| 13 | *1* | **220** | *14* | **202** | *7* | **175** | *12* | **205** |
| 14 | *13* | **205** | *2* | **152** | *11* | **180** | *8* | **187** |
| 15 | *5* | **165** | *10* | **150** | *3* | **200** | *16* | **160** |
| 16 | *9* | **155** | *6* | **177** | *15* | **185** | *4* | **172** |
| | | | | REPLICATION V | | | | |
| 17 | *1* | **147** | *10* | **112** | *15* | **177** | *8* | **147** |
| 18 | *9* | **180** | *2* | **205** | *7* | **190** | *16* | **167** |
| 19 | *13* | **172** | *6* | **212** | *3* | **197** | *12* | **192** |
| 20 | *5* | **177** | *14* | **220** | *11* | **205** | *4* | **225** |

* Fertilizer treatment number

† **Tiller number/m²**

**Table 5.3** Source of variation and degrees of freedom for a balanced lattice design experiment

| SOURCE OF VARIATION | DEGREES OF FREEDOM (DF) | RESULTS OF DF |
|---|---|---|
| Replication (rep) | $(k+1)-1$ | $(4+1)-1=4$ |
| Treatments, unadjusted (trt) | $k^2-1$ | $4^2-1=15$ |
| Block, adjusted (block\|rep) | $(k+1)(k-1)$ | $(4+1)(4-1)=15$ |
| Intrablock error (Residual) | $(k-1)(k^2-1)$ | $(4-1)(4^2-1)=45$ |
| Treatment, adjusted | $k^2-1$ | $4^2-1=15$ |
| Effective error | $(k-1)(k^2-1)$ | $(4-1)(4^2-1)=45$ |

*Note:* Arrows indicate the ratio of mean squares for calculating F values.

Enter the following command in Stata:

```
anova tiller rep trt block|rep, sequential
```

As mentioned with the BIB design, the order the variables are entered and the option sequential are important for the design. The results of this command are

```
      Number of obs =        80     R-squared     =  0.7531
      Root MSE      = 17.9712       Adj R-squared =  0.5665

Source     |    Seq. SS     df        MS             F      Prob > F
-----------+----------------------------------------------------------
     Model |  44322.2375    34   1303.59522        4.04       0.0000
           |
       rep |     5946.05     4    1486.5125        4.60       0.0034
       trt |    26994.35    15   1799.62333        5.57       0.0000
 block|rep |  11381.8375    15   758.789167        2.35       0.0138
           |
  Residual |  14533.3125    45     322.9625
-----------+----------------------------------------------------------
     Total |    58855.55    79   745.006962
```

Look at the ANOVA table comparing the intrablock mean square (Residual) to the block, adjusted (block|rep) mean square. If the intrablock mean square is larger than the block, adjusted mean square, no further calculations are required and the results presented in the above ANOVA table are correct (see Gomez and Gomez, 1984). The analysis is not complete, in this case, however, because the intrablock mean square (322.9625) is less than the block, adjusted mean square (758.789167).

At this point an adjustment term must be calculated as well as adjustments to the treatment means. To calculate the adjustment term is rather tedious, so a Do-File has been included that does this. Open the Do-File ballatadj.do by selecting the Do… menu item under the File menu. The ballatadj.do file is in the Do-Files folder that is available with the book. Once the ballatadj.do file is open, run the file by selecting the Run icon in the upper right corner of the Do-File Editor. Once this file has been run, enter the following command:

```
ballatadj tiller rep trt block
```

which results in the following output:

```
         Number of obs =        80     R-squared     =  0.7531
         Root MSE      = 17.9712     Adj R-squared =  0.5665

  Source |    Seq. SS      df        MS           F     Prob > F
---------+------------------------------------------------------
   Model |  44322.2375     34  1303.59522       4.04      0.0000
         |
     rep |     5946.05      4   1486.5125       4.60      0.0034
     trt |    26994.35     15  1799.62333       5.57      0.0000
   block |  11381.8375     15  758.789167       2.35      0.0138
         |
Residual |  14533.3125     45    322.9625
---------+------------------------------------------------------
   Total |    58855.55     79  745.006962

Balanced Lattice Design with Adjustments

Treatment (adj.) MS:                 1600.116667
Effective error (residual) MS:       369.3375921
Computed F:                          4.332395892
Prob > F:                                 0.0001
Coefficient of Variation:                  11.2%
Relative Efficiency over an RCB:             17%
```

The adjustment did not result in a significantly different result from the original analysis, but this will not always be the case. In addition, the coefficient of variation (CV) and the relative efficiency compared to the RCBD are calculated. The specifics of the calculations of this Do-File are presented in the Appendix. Gomez and Gomez (1984) have a good presentation of this analysis.

## Group Balanced Block Design

In the previous section, to help control variability, a new factor was introduced, the block, which helps control variability in the experiment due to field position. The group balanced block design attempts to control variability by identifying a factor associated with the treatments themselves. This design may be used with large variety trials where, for example, maturity class or growth habit may be distinctive among the varieties.

The design is arranged much like an RCBD with the difference that the treatments are randomized within groups in each replication. So, for example, in a trial of 45 varieties with 3 groups of 15 varieties of different maturity, the varieties would be randomized within each group within a replication. Because of the way the experiment is

**Table 5.4** Source of variation and degrees of freedom for a group balanced block design experiment

| SOURCE OF VARIATION | DEGREES OF FREEDOM (DF) | RESULTS OF DF |
|---|---|---|
| Replication (rep) | $r-1$ | $3-1=2$ |
| Group (maturity) | $g-1$ ← | $3-1=2$ |
| Replication × Group (rep#maturity) | $(r-1)(g-1)$ | $(3-1)(3-1)=4$ |
| Treatment with Group 1 | $\frac{t}{g}-1$ ← | $\frac{45}{3}-1=14$ |
| Treatment with Group 2 | $\frac{t}{g}-1$ ← | $\frac{45}{3}-1=14$ |
| Treatment with Group 3 | $\frac{t}{g}-1$ ← | $\frac{45}{3}-1=14$ |
| Error (Residual) | $g(r-1)\left(\frac{t}{g}-1\right)$ | $3(3-1)\left(\frac{45}{3}-1\right)$ |

arranged, treatments within a group can be compared to each other with a greater degree of precision than treatments in different groups. Table 5.4 shows the degrees of freedom for an experiment with 45 varieties arranged into 3 groups of 15 varieties.

Load the dataset GroupBalBlock.dta into Stata. This dataset is from a variety trial of 45 rice varieties that consisted of 15 varieties in each of 3 different maturity groups (Gomez and Gomez, 1984, p. 77). The groups have maturities of less than 105 days, 105–115 days, and those that mature in over 115 days. After loading the dataset, enter the following command:

```
anova yield rep maturity/maturity#rep var|maturity
```

This results in the following output:

```
Number of obs =      135     R-squared     =  0.7485
Root MSE      = .296459      Adj R-squared =  0.5988

      Source |  Partial SS    df       MS          F     Prob > F
 ------------+----------------------------------------------------
       Model |  21.9712995    50   .43942599     5.00     0.0000
             |
         rep |  5.52888354     2  2.76444177    17.48     0.0105
    maturity |  3.35749913     2  1.67874957    10.61     0.0251
maturity#rep |  .632773299     4  .158193325
 ------------+----------------------------------------------------
```

```
var|maturity |  12.4521435    42  .296479608    3.37      0.0000
             |
    Residual |  7.38259695    84  .087888059
-------------+----------------------------------------------------
  Total |  29.3538965   134  .219058929
```

The maturity groups are clearly different with an F value of 10.61. The maturity by replication interaction (`maturity#rep`, 0.158193325) is the mean square error used as the denominator to calculate this value. This is accomplished in the command by using the / character between `maturity` and `rep#maturity`.

The variety within maturity group sum of squares (12.4521435) needs to be partitioned for each maturity group so that an accurate F value can be calculated for each. The residual mean square (0.087888059) is the correct term to use for the denominator in calculating these F values, so at this point this value should be stored in a macro. Enter the following command:

```
local x = e(rmse)^2
local y = e(df_r)
```

Remember the Root MSE (0.296459) is the square root of the Residual mean square (0.087888059), which is the value we are interested in saving for future calculations. In addition, we are interested in saving the Residual degrees of freedom e(df_r), which in this case is 84. Now, enter the following command:

```
anova yield var rep if maturity == 1
```

This calculates an ANOVA based on the first maturity class and results in the following output:

```
     Number of obs =        45     R-squared     =  0.7668
     Root MSE      =  .262096     Adj R-squared =  0.6336

  Source |  Partial SS    df       MS           F     Prob > F
  -------+----------------------------------------------------
   Model |   6.3255123    16  .395344519       5.76     0.0000
         |
     var |  4.15479519    14  .296771085       4.32     0.0005
     rep |  2.17071711     2  1.08535855      15.80     0.0000
         |
Residual |  1.92344273    28  .068694383
---------+----------------------------------------------------
   Total |  8.24895503    44  .187476251
```

The Residual mean square from the previous ANOVA is used to calculate the correct F value, so the following commands are entered:

```
local z = e(df_1)
local g1 = e(ss_1)/e(df_1)/`x'
display `g1'
display Ftail(`z', `y', `g1')
```

The first command stores the var degrees of freedom, which is 14 in this case in the macro z. The second command calculates the correct F value for the first group. The first part of this equation, e(ss_1)/e(df_1), divides the Partial SS (4.15479519) by the degrees of freedom (14) to calculate the var mean square (0.296771085), which is then divided by x, the mean square from the previous ANOVA. The next command displays the results of this calculation (3.376694). Finally, the last command calculates the probability associated with the numerator and denominator degrees of freedom with this F value and displays the results (0.00025122).

The above shows how the variety mean square is partitioned and divided by the residual mean square from the overall ANOVA. This can, however, be easily handled by Stata with the following command:

```
contrast var|maturity
```

which results in the following output:

```
Contrasts of marginal linear predictions

Margins      : asbalanced

------------------------------------------------
             |         df           F        P>F
-------------+----------------------------------
var|maturity |
          1  |         14        3.38     0.0003
          2  |         14        2.11     0.0192
          3  |         14        4.64     0.0000
      Joint  |         42        3.37     0.0000
             |
   Residual  |         84
------------------------------------------------
```

This command must be entered immediately after the estimation command:

```
anova yield rep maturity/maturity#rep var|maturity
```

**Subsampling**

Oftentimes it is desirable or necessary to collect subsamples from within experimental units. This introduces another source of variability often called *sampling error*. Such sampling may be desirable particularly with items that can be easily measured or that are prone to a great deal of variability. For example, plant height might be better represented with several measurements rather than a single plant within an experimental unit, while measuring every plant in the experimental unit may be too time consuming or costly.

Open the dataset Watermelon Subsampling.dta. This is a dataset from a variety trial where two fruits from each experimental unit were measured for length, width, rind thickness, and percent soluble solids (sugar content). In addition, there is a variable representing the ratio of the length to width (`lwratio`). Enter the following command and see the results:

```
anova lwratio rep trt/rep#trt

                 Number of obs =       168     R-squared     =  0.8887
                 Root MSE      = .130343       Adj R-squared =  0.7788

      Source |  Partial SS    df       MS           F     Prob > F
  -----------+----------------------------------------------------
       Model |  11.3971347    83  .137314876       8.08     0.0000
             |
         rep |  .203931699     3  .067977233       1.04     0.3798
         trt |  7.28573298    20  .364286649       5.59     0.0000
     rep#trt |  3.90747006    60  .065124501
  -----------+----------------------------------------------------
             |
    Residual |  1.42711017    84  .016989407
  -----------+----------------------------------------------------
       Total |  12.8242449   167  .076791886
```

The experimental error term (denominator) to calculate the F-test, in this case, is the replication by treatment interaction (`rep#trt`). In an RCBD without subsampling, the experimental error term would simply be the residual. Table 5.5 shows the correct terms to use in calculating a CRD, RCBD, and a split-plot design with subsampling. The important thing to note is that sampling error has been accounted for and that the appropriate error term is used.

**Table 5.5** Source of variation and degrees of freedom for CRD, RCBD, and split-plot designs with subsampling

| | DEGREES OF FREEDOM (DF) | | |
|---|---|---|---|
| SOURCE OF VARIATION | CRD | RCBD | SPLIT-PLOT DESIGN |
| Replication (r) | $r-1$ | $r-1$ | $r-1$ |
| Main-plot treatment (a) | $a-1$ ← | $a-1$ ← | $a-1$ ← |
| Error | $a(r-1)$ | $(r-1)(t-1)$ | $(r-1)(t-1)$ |
| Subplot treatment (b) | | | $b-1$ ← |
| a × b | | | $(a-1)(b-1)$ ← |
| Error | | | $a(r-1)(b-1)$ |
| Sampling error (s) | $rt(s-1)$ | $rt(s-1)$ | $abr(s-1)$ |

*Note:* Arrows indicate the ratio of mean squares for calculating F values.

Several criteria should be considered with subsampling. It should be easy to obtain, have good precision, and be low cost. Subsampling information from previous experiments also can help determine sample size for future experiments.

One approach evaluates the variance of a treatment mean and the CV to determine an appropriate subsample size. Deciding on a sample size should have a low sampling variance and meet the degree of precision desired. Computing the variance of a treatment mean can be accomplished by calculating the experimental error variance as follows:

$$\sigma_e^2 = \frac{\sigma_{e+s}^2 - \sigma_s^2}{n}$$

where the $\sigma_{e+s}^2$ is the mean square for `rep#trt` and $\sigma_s^2$ is the residual mean square. The $n$ is the number of subsamples. To do this from the above ANOVA, enter the following:

```
display (.065124501-.016989407)/2
```

which results in 0.02406755. The experimental error variance is then used to calculate the variance of a treatment mean represented by the formula

$$\sigma^2 = \frac{\sigma_s^2 + n\sigma_e^2}{rn}$$

where the variables are defined above and $r$ is the number of replications. Enter the following and see the results:

```
display (.016989407+(2*.02406755))/(4*2)
```

The result is 0.00814056.

Next we wish to calculate the CV or standard error of the treatment mean expressed as a percent. This formula is

$$CV = \frac{100\sqrt{\sigma^2}}{\bar{X}}$$

The overall mean ($\bar{X}$) is required to calculate the CV and can be acquired with the command

```
summarize lwratio
```

The **summarize** command computes several statistics including the mean, which is 1.307622. Finally, to calculate the CV, enter

```
display 100*sqrt(.00814056)/1.307622
```

This calculates to 6.8999338%. At this point it is possible to substitute different numbers of subsamples or replications in the above formulas to see what effect it has on the CV and, thus, the precision of the experiment. Increasing either subsamples or replications will lower the CV. The right combination should have a reasonable CV while not consuming too many resources. In this case, if the number of subsamples is increased to 10, the CV is only reduced from the original 6.9% to 6.1%. Adding one more replication also only reduces the CV to 6.2%.

Using a somewhat different approach with the above information, it can be used to estimate the number of subsamples based on the level of significance required and the margin of error as a fraction of the treatment mean. The following formula can be used to calculate this value:

$$n = \frac{\left(Z_a^2\right)\left(\sigma_s^2\right)}{r\left(D^2\right)\left(\bar{X}^2\right) - \left(Z_a^2\right)\left(\sigma_e^2\right)}$$

where $Z_a$ is the standard normal density, $\sigma_s^2$ is the subsample variance, $r$ is the number of replications, $D$ is the margin of error as a decimal percent of the treatment mean, and $\sigma_e^2$ is the experimental error variance.

Stata can calculate the number of subsamples with this formula with the following input:

```
display (invnormal(0.025)^2*.016989407)/
(4*.05^2*1.307622)-(invnormal(0.025)^2*.02406755)
```

It is common to use a value of 1.96 for the standard normal density ($Z_a$) since this represents the 0.05 probability for this function. Stata can calculate this density function with the **invnormal** command. The value entered is 0.025 (half of 0.05) because it only calculates half the function; therefore, by using this value, it gives the correct value, 1.96. This formula results in 4.8985985 or that 5 subsamples should be taken to meet the criteria. If the margin of error were raised from 0.05 to 0.1, it would result in 1.1553088 or that 2 subsamples should be taken.

# 6

# Two and More Factors ANOVA

Up until this point, all of the experiments (utilizing ANOVA [analysis of variance]) we have examined have dealt with a single experimental factor whether varieties, fertility levels, etc. It is possible to conduct experiments in which more than one experimental factor at a time is considered. In fact, in most biological systems, there are several factors in play at any one time. Conducting experiments with more than one factor is more likely to mimic real environmental conditions and allows the researcher to see how these factors interact with one another. In addition, time and resources may be conserved because more than one factor is considered in a single experiment.

The number of factors that can be considered in an experiment is theoretically unlimited. In fact, the analysis of experiments with up to five factors has been worked out with considerations of various combinations of random and fixed effect models. Caution, however, should be exercised when considering experiments with greater than three factors. It has been shown that a random set of numbers coded with a high number of factors will have a high likelihood of showing some significance.

Factorial experiments can be implemented in any number of experimental designs, such as RCBD (randomized complete block design), split-plot design, split-block design, etc. In addition, factorial experiments can include more than two factors; however, the number of factors is usually limited because of resource limitations and the potential increase in type I errors. Figure 6.1 shows several possible interaction effects that might occur in an experiment. Statistical analysis of factorial experiments can help identify these interactions and help determine why they occur.

Load the dataset SeedstemFactor.dta. This experiment is an example of a factorial experiment arranged as an RCBD involving onion varieties and sowing dates as the treatment factors. The dataset is of

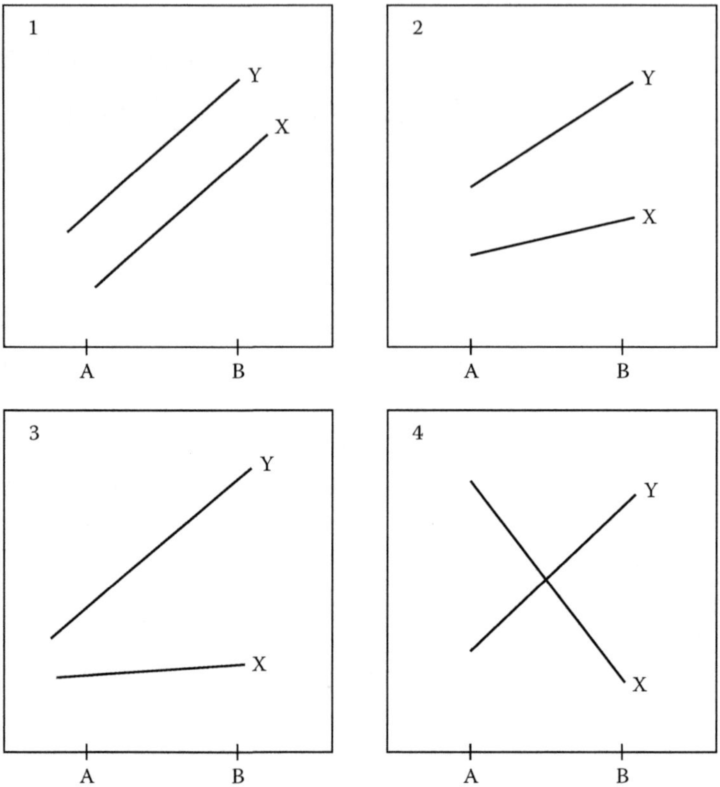


**Figure 6.1** Various interactions between factor AB and factor XY. 1 = no interaction, 2 = low change in magnitude effect, 3 = high change in magnitude effect, 4 = strong interaction effect.

onion seedstems or flowers. Flowering in an onion production crop is considered an undesirable characteristic because such onions are culled, which reduces yield. Enter the following command:

```
anova seedstem rep variety date variety#date
```

This results in the following output:

```
Number of obs =        48     R-squared     =  0.8071
Root MSE      = 5.06648       Adj R-squared =  0.7253

      Source |  Partial SS    df       MS          F      Prob > F
-------------+----------------------------------------------------
       Model |  3544.16667    14  253.154762     9.86      0.0000
             |
         rep |  77.4166667     3  25.8055556     1.01      0.4027
```

```
     variety |     741.75      3      247.25    9.63      0.0001
        date |   1877.625      2    938.8125   36.57      0.0000
variety#date |    847.375      6  141.229167    5.50      0.0005
             |
    Residual | 847.083333     33  25.6691919
-------------+----------------------------------------------------
       Total |    4391.25     47  93.4308511
```

When dealing with two factors (in this case, varieties and sowing dates) to show the interaction between these terms, the **#** character is used between these two factors to calculate the interaction effect. All three—`variety`, `date`, and `variety#date`—are significant. At this point, you would want to explore this interaction further. To do this, you may wish to look at the mean seedstem values for the varieties and dates. Enter the following command:

```
table variety date, contents(mean seedstem)
```

This results in the following output:

```
--------------------------------------
Varieties:     |
1-Pegasus,     |
2-Swt.         |
Vidalia,       | Sowing date: 1-5 Oct,
3-Nirvana,     |   2-15 Oct, 3-29 Oct
4-PS 7092      |  5 Oct  15 Oct  29 Oct
---------------+----------------------
Pegasus        |   9.25     .25       0
Sweet Vidalia  |  30.75       5       0
Nirvana        |      7       0     .25
PS 7092        |   8.75     .25       0
--------------------------------------
```

This shows that, as the sowing date gets later, there appears to be a reduction in the number of seedstems. In addition, Sweet Vidalia appears to have more seedstems than the other varieties, at least for the first and second sowing dates. This can be explored further by examining ANOVA tables for varieties over the different sowing dates as well as evaluating ANOVA tables for the different sowing dates for each variety. For the former case, enter the following command:

```
by date, sort : anova seedstem rep variety
```

This results in the following three ANOVA tables, one for each sowing date.

```
----------------------------------------------------------------
-> date = 5 Oct

          Number of obs =        16     R-squared     =  0.8402
          Root MSE      = 6.39499       Adj R-squared =  0.7336

       Source | Partial SS     df        MS            F     Prob > F
   -----------+----------------------------------------------------
        Model |   1934.875      6  322.479167        7.89     0.0035
              |
          rep |   416.1875      3  138.729167        3.39     0.0674
      variety |  1518.6875      3  506.229167       12.38     0.0015
              |
     Residual |   368.0625      9  40.8958333
   -----------+----------------------------------------------------
        Total |  2302.9375     15  153.529167

----------------------------------------------------------------
-> date = 15 Oct

          Number of obs =        16     R-squared     =  0.4839
          Root MSE      = 3.46811       Adj R-squared =  0.1398

       Source | Partial SS     df        MS            F     Prob > F
   -----------+----------------------------------------------------
        Model |      101.5      6  16.9166667        1.41     0.3096
              |
          rep |      31.25      3  10.4166667        0.87     0.4934
      variety |      70.25      3  23.4166667        1.95     0.1927
              |
     Residual |     108.25      9  12.0277778
   -----------+----------------------------------------------------
        Total |     209.75     15  13.9833333

----------------------------------------------------------------
-> date = 29 Oct

          Number of obs =        16     R-squared     =  0.4000
          Root MSE      =       .25     Adj R-squared =  0.0000

       Source | Partial SS     df        MS            F     Prob > F
   -----------+----------------------------------------------------
        Model |       .375      6       .0625        1.00     0.4799
              |
          rep |      .1875      3       .0625        1.00     0.4363
      variety |      .1875      3       .0625        1.00     0.4363
              |
     Residual |      .5625      9       .0625
   -----------+----------------------------------------------------
        Total |      .9375     15       .0625
```

The differences between the varieties occur only with the first sowing date with a significant difference between the varieties (p = 0.0015), whereas there isn't any difference between the varieties on the second and third sowing dates.

Another way to view these data is with the **contrast** command. This command allows you to view any linear hypothesis involving factor variables and their interactions. Enter the following command immediately after the factorial ANOVA:

```
contrast variety@date
```

which results in the following output:

```
Contrasts of marginal linear predictions

Margins      : asbalanced

------------------------------------------------
             |         df           F        P>F
-------------+----------------------------------
variety@date |
          1  |          3       19.72     0.0000
          2  |          3        0.91     0.4457
          3  |          3        0.00     0.9998
      Joint  |          9        6.88     0.0000
             |
   Residual  |         33
------------------------------------------------
```

This command calculates the probabilities of seedstem differences between the varieties for each of the sowing dates. Notice the F values are different from the ANOVA tables calculated for each date above. This is because these F values are calculated using the mean square for the residuals (25.6691919) from the overall ANOVA table as the denominator rather than the mean square for the residuals from the individual ANOVA tables by date. The overall mean square for the residuals is a more appropriate denominator if the individual residual mean squares are all similar. This overall residual mean square is based on larger degrees of freedom, therefore, it will be a smaller value than for the individual ANOVAs, which means it has greater power to detect differences. If, however, the individual ANOVAs have residual mean squares that are quite different from each other, as in this case, then the individual ANOVAs would be more appropriate.

In addition to looking at the variety ANOVA tables individually, the sowing dates also can be examined for each variety with the following command:

```
contrast date@variety
```

This results in the following output:

```
Contrasts of marginal linear predictions

Margins      : asbalanced

------------------------------------------------
             |         df           F        P>F
-------------+----------------------------------
date@variety |
          1  |          2        4.33     0.0214
          2  |          2       42.43     0.0000
          3  |          2        2.46     0.1012
          4  |          2        3.87     0.0310
      Joint  |          8       13.27     0.0000
             |
   Residual  |         33
------------------------------------------------
```

With the exception of variety 3 (Nirvana), the other varieties have significantly lower seedstem numbers with later sowing dates at the p = 0.05 level of significance. You may wish to try calculating the individual ANOVAs to see how they differ from these results.

### Split-Plot Design

A split-plot design is another type of factorial design usually used because of some limitation in space or to facilitate treatment application. The two factors are divided into a main-plot effect and a subplot effect. The precision is greater for the subplot factor than it is for the main-plot factor. If one factor is more important to the researcher and if the experiment can facilitate it, then the subplot factor should be used for this factor. This may not always be the case, however.

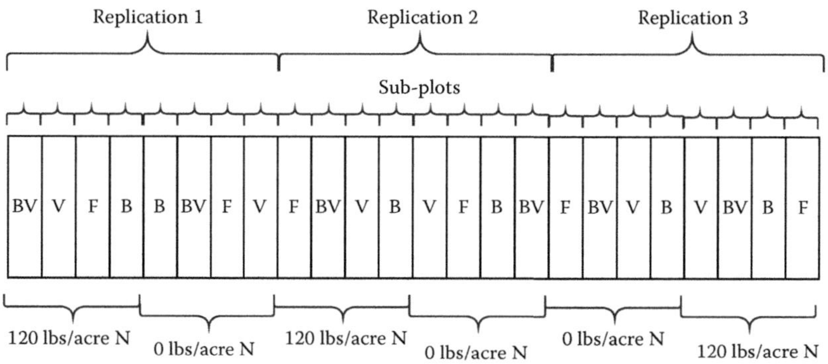


**Figure 6.2** Layout of a split-plot design. Main plots are different fertilizer rates (0 or 120 lbs/acre nitrogen). Subplots are green manures. BV = barley–vetch, V = vetch, F = fallow, and B = barley. (From Little, T. M., and F. J. Hills. 1978. *Agricultural Experimentation Design and Analysis*. New York: John Wiley & Sons, p. 89. With permission.)

**Table 6.1** Source of variation and degrees of freedom for a split-plot design experiment

| SOURCE OF VARIATION | DEGREES OF FREEDOM (DF) | RESULTS OF DF |
|---|---|---|
| Replication (rep) | $r-1$ | $3-1=2$ |
| Main plot (fert) | $a-1$ ← | $2-1=1$ |
| Main-plot error (rep#fert) | $(r-1)(a-1)$ | $(3-1)(2-1)=2$ |
| Sub-plot (green) | $b-1$ ← | $4-1=3$ |
| Main-plot x sub-plot interaction (fert#green) | $(a-1)(b-1)$ ← | $(2-1)(4-1)=3$ |
| Sub-plot error (Residual) | $a(r-1)(b-1)$ | $2(3-1)(4-1)=12$ |

*Note:* Arrows indicate the ratio of mean squares for calculating F values.

Figure 6.2 shows the layout of a split-plot design with fertilizer rates as the main-plot effects (0 or 120 lbs/acre N) and the subplot affects green manure effects (barley–vetch, vetch, fallow, or barley) as the subplot effects. As mentioned previously, the level of precision will be different for the main plots compared to the subplots and this has to do with which value is used in the denominator to determine the F value for each factor. Table 6.1 shows the degrees of freedom with the arrows indicating the devisors for each factor.

Load the dataset 'Factorial.dta', (Little and Hill, 1978, p. 90) and enter the following command:

```
anova yield rep fert/rep#fert green fert#green
```

This results in the following output:

```
Number of obs    =      24      R-squared      =  0.9860
  Root MSE       = .776656      Adj R-squared =  0.9731

     Source |  Partial SS    df       MS           F     Prob > F
  ----------+----------------------------------------------------
      Model |  508.881253    11   46.261932      76.69    0.0000
            |
        rep |  7.86583229     2  3.93291615       1.56    0.3903
       fert |  262.020418     1  262.020418     104.06    0.0095
   rep#fert |   5.0358341     2  2.51791705
  ----------+----------------------------------------------------
      green |  215.261242     3  71.7537472     118.96    0.0000
 fert#green |  18.6979266     3  6.23264219      10.33    0.0012
            |
   Residual |   7.2383355    12  .603194625
  ----------+----------------------------------------------------
      Total |  516.119588    23  22.4399821
```

Both the fertilizer and green manure, as well as the fertilizer/green manure, interactions were significant. To examine these values more closely and determine what the effects are, use the **margins** and **marginsplot** commands. The **margins** command in this case can calculate the means for all combinations of fertilizer and green manure and must be entered following an estimation command, such as **anova** used here. To do this, enter the command

```
margins fert#green
```

This results in the following output:

```
Predictive margins                     Number of obs    =         24

Expression   : Linear prediction, predict()

------------------------------------------------------------------
            |       Delta-method
            | Margin   Std. Err.   z     P>|z|  [95% Conf. Interval]
------------+-----------------------------------------------------
fert#green  |
   0 1      |     13.5 .4484026  30.11 0.000  12.62115    14.37885
   0 2      |15.23333 .4484026  33.97 0.000  14.35448    16.11219
   0 3      |       22 .4484026  49.06 0.000  21.12115    22.87885
   0 4      |18.93333 .4484026  42.22 0.000  18.05448    19.81219
 120 1      |19.26667 .4484026  42.97 0.000  18.38781    20.14552
 120 2      |23.93333 .4484026  53.37 0.000  23.05448    24.81219
 120 3      |26.16667 .4484026  58.36 0.000  25.28781    27.04552
 120 4      |26.73333 .4484026  59.62 0.000  25.85448    27.61219
------------------------------------------------------------------
```

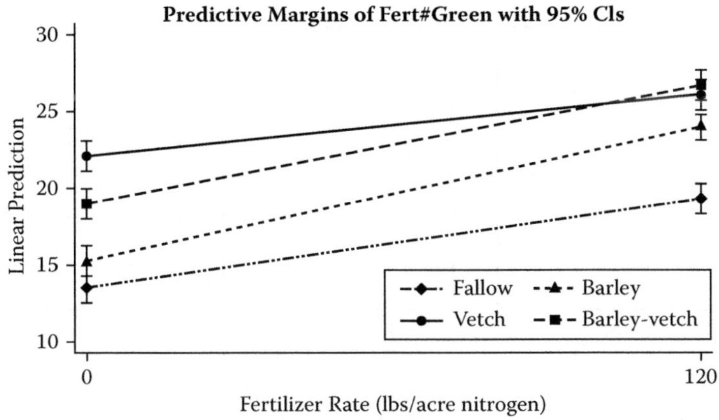


**Figure 6.3** Output from **marginplots** command showing interactions of fertilizer rates and green manure effects on sugar beet yields (tons/acre).

The Margin column lists the means for each combination of fertilizer and green manure. This will not always be the case as we will see in covariance analysis. After this command is entered, the **`marginsplot`** command can be entered (this command must follow the **`margins`** command), which results in the graph shown in Figure 6.3.

Both the fertilizer and green manure treatments had an effect on sugar beet yields. There also was an interaction effect between the two factors. An examination of the green manure effects with and without fertilizer indicated treatments with vetch (vetch or vetch–barley) appeared to have higher yields than green manures without vetch regardless of fertilizer application. Fertilizer also had a significant effect on yield for all green manures. The fertilizer effect, however, was greater with barley and barley–vetch than with vetch alone or for the fallow treatment.

### Split-Block Design

The split-block design, which is also referred to as a strip-plot design, is a derivation of the split-plot design. In this design, the first factor is randomly assigned to plots in one direction and the second factor is randomly assigned perpendicular to the first factor. This type of design is often used where treatment application is applied by equipment (e.g., fertilizer or herbicide application equipment). Treatments are applied

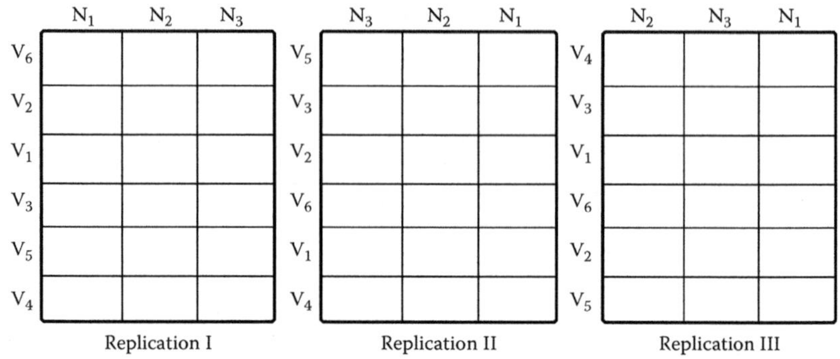


**Figure 6.4** Layout of a split-block design. Horizontal treatments are six different rice varieties and vertical treatments are three different nitrogen rates (0, 60, 120 kg/ha). (From Gomez, K. A., and A. A. Gomez. 1984. *Statistical Procedures for Agricultural Research*, 2nd ed. New York: John Wiley & Sons, p. 110.)

in a continuous strip, which is easier when using equipment. Figure 6.4 shows an example of just such a layout where the horizontal treatments are varieties and the vertical treatments are nitrogen fertilizer rates. It is easy to see, particularly for the fertilizer application, that treatment application is more easily facilitated with such a design.

The degree of precision for measuring the two factors is equivalent, while the degree of precision for the interaction effect is increased (Table 6.2). Along with facilitating treatment application, such designs would be desirable where the interaction effect is of particular interest.

Load the dataset Splitblock.dta, which is a dataset of six rice varieties and three nitrogen fertilizer rates (Gomez and Gomez, 1984, p. 110). Enter the following command:

```
anova yield rep var/var#rep fert/fert#rep var#fert/
rep#var#fert
```

This will result in the following output:

```
Number of obs =        54     R-squared     =  1.0000
Root MSE      =         0     Adj R-squared =

      Source |  Partial SS     df       MS             F     Prob > F
-------------+----------------------------------------------------
       Model |   167005649     53  3151049.98
             |
         rep |  9220962.33      2  4610481.17        3.09     0.0902
         var |  57100201.3      5  11420040.3        7.65     0.0034
     var#rep |  14922619.2     10  1492261.92
```

**Table 6.2** Source of variation and degrees of freedom for a split-block design experiment

| SOURCE OF VARIATION | DEGREES OF FREEDOM (DF) | RESULTS OF DF |
|---|---|---|
| Replication (`rep`) | $r-1$ | $3-1=2$ |
| Horizontal factor (`var`) | $a-1$ ← | $6-1=5$ |
| Horizontal factor error (`rep#var`) | $(r-1)(a-1)$ | $(3-1)(6-1)=10$ |
| Vertical factor (`fert`) | $b-1$ ← | $3-1=2$ |
| Vertical factor error (`rep#fert`) | $(r-1)(b-1)$ | $(3-1)(3-1)=4$ |
| Variety × fertilizer interaction (`var#fert`) | $(a-1)(b-1)$ ← | $(6-1)(3-1)=10$ |
| Variety × fertilizer error (`rep#var#fert`) | $(r-1)(a-1)(b-1)$ | $(3-1)(6-1)(3-1)=20$ |

*Note:* Arrows indicate the ratio of mean squares for calculating F values.

```
-------------+----------------------------------------------------
        fert |  50676061.4      2  25338030.7       34.07     0.0031
    fert#rep |  2974907.89      4  743726.972
-------------+----------------------------------------------------
    var#fert |  23877979.4     10  2387797.94        5.80     0.0004
rep#var#fert |  8232917.22     20  411645.861
-------------+----------------------------------------------------
             |
    Residual |           0      0
-------------+----------------------------------------------------
       Total |   167005649     53  3151049.98
```

Looking at the results, we see that variety (`var`) and fertility (`fert`) rates are significant. In addition, the variety by fertility interaction is significant as well.

Because the fertilizer was applied at equally spaced rates, it is possible to examine this factor as a linear effect (regression and correlation will be discussed more fully in Chapter 10). Examine the dataset and you will see the fertilizer rates are entered as they were applied: 0, 60, and 120 kg/ha. Entering a c. prior to a variable tells Stata to treat this variable as continuous rather than as discrete values. Enter the following command:

```
anova yield rep var/var#rep c.fert/c.fert#rep
var#c.fert/rep#var#c.fert
```

This results in the following output:

```
Number of obs =        54     R-squared     =  0.9426
Root MSE      = 729.623       Adj R-squared =  0.8311

     Source  |  Partial SS    df       MS           F     Prob > F
 ------------+-------------------------------------------------------
      Model  |   157423365    35  4497810.44      8.45     0.0000
             |
        rep  |  9906491.23     2  4953245.62      5.83     0.0210
        var  |  7133423.66     5  1426684.73      1.68     0.2271
    var#rep  |  8500280.86    10  850028.086
 ------------+-------------------------------------------------------
       fert  |  49718951.4     1  49718951.4     38.51     0.0250
   fert#rep  |  2582254.39     2  1291127.19
 ------------+-------------------------------------------------------
   var*fert  |  21478590.5     5  4295718.09     17.90     0.0001
rep*var*fert|  2399786.28    10  239978.628
 ------------+-------------------------------------------------------
             |
   Residual  |   9582283.5    18  532349.083
 ------------+-------------------------------------------------------
      Total  |   167005649    53  3151049.98
```

In this analysis, as in the previous, the fertilizer rate is significant (Prob>F) at 0.0250. This also tells us, however, that the effect of the fertilizer application was a linear effect.

Another approach is to drop the last term and do the analysis again. Enter the following and see the results:

**anova** yield rep var/var#rep fert/fert#rep var#fert

```
Number of obs =        54     R-squared     =  0.9507
Root MSE      = 641.596       Adj R-squared =  0.8694

     Source  |  Partial SS    df       MS           F     Prob > F
 ------------+-------------------------------------------------------
      Model  |   158772732    33  4811294.9      11.69     0.0000
             |
        rep  |  9220962.33     2  4610481.17      3.09     0.0902
        var  |  57100201.3     5  11420040.3      7.65     0.0034
    var#rep  |  14922619.2    10  1492261.92
 ------------+-------------------------------------------------------
       fert  |  50676061.4     2  25338030.7     34.07     0.0031
   fert#rep  |  2974907.89     4  743726.972
 ------------+-------------------------------------------------------
   var#fert  |  23877979.4    10  2387797.94      5.80     0.0004
             |
   Residual  |  8232917.22    20  411645.861
 ------------+-------------------------------------------------------
      Total  |   167005649    53  3151049.98
```

The results are essentially the same, but now we can use the **margins** and **marginsplot** commands to examine the var#fert interaction. Enter the following command and see the results:

```
margins var#fert

Predictive margins                   Number of obs    =         54

Expression    : Linear prediction, predict()

------------------------------------------------------------------------
         |         Delta-method
         | Margin   Std. Err.     z     P>|z|      [95% Conf. Interval]
---------+--------------------------------------------------------------
var#fert |
 1    0  |3571.667   370.4258   9.64   0.000      2845.645    4297.688
 1   60  |    5132   370.4258  13.85   0.000      4405.979    5858.021
 1  120  |    7548   370.4258  20.38   0.000      6821.979    8274.021
 2    0  |4934.333   370.4258  13.32   0.000      4208.312    5660.355
 2   60  |6713.667   370.4258  18.12   0.000      5987.645    7439.688
 2  120  |7211.333   370.4258  19.47   0.000      6485.312    7937.355
 3    0  |4249.667   370.4258  11.47   0.000      3523.645    4975.688
 3   60  |6122.333   370.4258  16.53   0.000      5396.312    6848.355
 3  120  |7868.333   370.4258  21.24   0.000      7142.312    8594.355
 4    0  |    4059   370.4258  10.96   0.000      3332.979    4785.021
 4   60  |5553.667   370.4258  14.99   0.000      4827.645    6279.688
 4  120  |7094.333   370.4258  19.15   0.000      6368.312    7820.355
 5    0  |4101.667   370.4258  11.07   0.000      3375.645    4827.688
 5   60  |5633.333   370.4258  15.21   0.000      4907.312    6359.355
 5  120  |    6012   370.4258  16.23   0.000      5285.979    6738.021
 6    0  |3207.333   370.4258   8.66   0.000      2481.312    3933.355
 6   60  |3714.333   370.4258  10.03   0.000      2988.312    4440.355
 6  120  |    2492   370.4258   6.73   0.000      1765.979    3218.021
------------------------------------------------------------------------
```

If you examine the Margin column, which contains the means sorted by variety and then fertilizer rate, you may begin to see the interaction. All the variety yields increase with increasing fertilizer application except for variety 6. To see this more clearly, enter the **margins** command as **margins** fert#var (this is in reverse order from above—results not shown) and then enter the **marginsplot** command. The results of the **marginsplot** are seen in Figure 6.5 and graphically show the differences between the varieties.

Another approach to see this interaction effect is to calculate a separate ANOVA for each variety. One way to do this is to examine the fertilizer rates effect for each variety. The following command will calculate an ANOVA for each variety:

```
by var, sort: anova yield fert rep
```

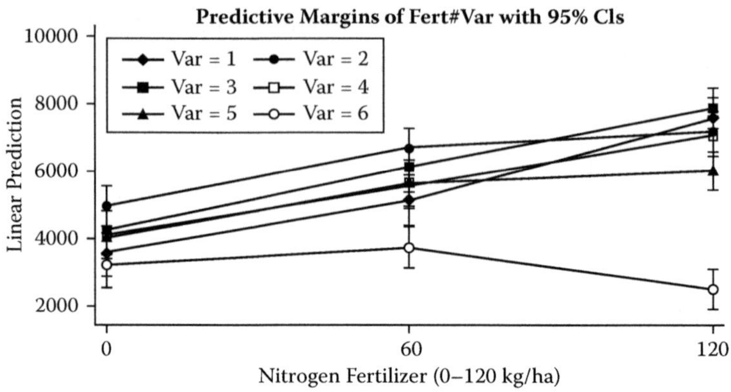


**Figure 6.5** Margin plots of rice varieties and the effect of nitrogen fertilizer.

The Prob > F values for the fertilizer rates for each variety were 0.0160, 0.0246, 0.0090, 0.0071, 0.0128, and 0.2210. In each case, as fertilizer was increased, there was an increase in yield with the exception of variety 6. This can be visually shown by entering the command

```
twoway lfit yield fert, by(var)
```

The **twoway** command, which is available under the Graphics menu, is one of the primary commands for displaying graphs. The **twoway lfit** command plots a linear prediction of the entered variables; in this case, yield and fertilization. The first variable (i.e., `yield`) is the y or ordinate variable and the second variable (i.e., `fert`) is the x or abscissa variable. (Graphing will be covered in more detail in Chapter 9.) The **by** modifier indicates plots should be drawn for each variety variable (i.e., `var`). See the output in Figure 6.6. Note how yield increases in each graph with increasing fertilization with the exception of variety 6. This is an example of how an interaction effect can affect results and elucidate a better understanding of the overall treatment effects.

### Evaluation over Years or Seasons

Evaluating data over years or seasons is a special case of a factorial design where years or seasons become a factor in the design. It is fairly

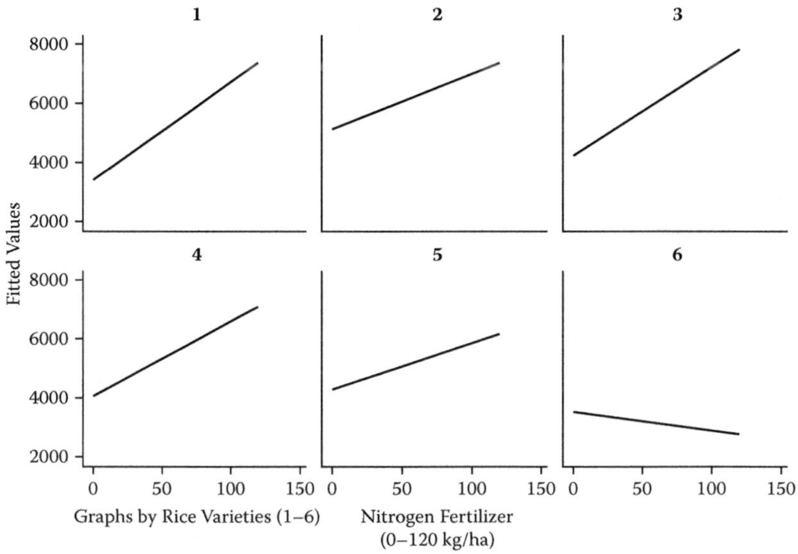


**Figure 6.6** Graphic output of fertilizer effect on yield for each of six rice varieties.

common and often required by refereed publications where two or more years of data are expected.

Evaluation over seasons is a case where the season variable is considered a fixed effect. In temperate climates, spring tends to be warming with increasing day length, while in the fall temperatures tend to fall and days get shorter. In tropical climates, the temperature or day length differences may not be that important, but there usually are seasonal differences often with wet and dry seasons. Although the specific conditions may change from year to year, the overall differences in seasons remain the same, or in statistical parlance, they are fixed (Table 6.3).

Open the file Pumpkin Seasons.dta, which is a dataset of variety trials conducted in the spring and again in the fall. Pumpkins are highly susceptible to a number of potyviruses that are particularly severe in the fall when aphid (insects that transmit the virus) populations peak. These trials were to evaluate a new variety, Orange Bulldog, which is resistant to many of these viruses. Enter the following command and see the results:

**Table 6.3** Source of variation and degrees of freedom for ANOVA for experiment over seasons and years

| SOURCE OF VARIATION | DEGREES OF FREEDOM (DF) | RESULTS OF DF |
|---|---|---|
| **ANALYSIS OF VARIANCE OVER SEASONS** | | |
| Seasons (`season`) | $s-1$ | $2-1=1$ |
| Replications within seasons (`repl season`) | $s(r-1)$ | $2(4-1)=6$ |
| Treatments (`var`) | $(t-1)$ | $7-1=6$ |
| Season x treatment (`season#var`) | $(s-1)(t-1)$ | $(2-1)(7-1)=6$ |
| Pooled error (`rep#varl season`) | $s(r-1)(t-1)$ | $2(4-1)(7-1)=36$ |
| **ANALYSIS OF VARIANCE OVER YEARS** | | |
| Years (`year`) | $y-1$ | $3-1=2$ |
| Replications within years (`repl year`) | $y(r-1)$ | $3(6-1)=15$ |
| Treatments (`var`) | $(t-1)$ | $8-1=7$ |
| Years × treatments (`year#var`) | $(y-1)(t-1)$ | $(3-1)(8-1)=14$ |
| Replications × treatment within years (`rep#varl year`) | $y(r-1)(t-1)$ | $3(6-1)(8-1)=105$ |

*Note:* Arrows indicate the ratio of mean squares for calculating F values.

```
anova wt season/rep|season var season#var/rep#var|season
```

```
           Number of obs =       84     R-squared     =  0.9634
           Root MSE      = 13.4639     Adj R-squared =  0.8916
   Source       |  Partial SS    df         MS           F    Prob > F
  --------------+----------------------------------------------------
          Model |  133692.817    55  2430.77849    13.41      0.0000
                |
         season |  812.240238     1  812.240238     2.14      0.1935
     rep|season |  2273.86821     6  378.978036
  --------------+----------------------------------------------------
            var |  106106.946     6  17684.4911    45.30      0.0000
     season#var |   5842.9131     6  973.818849     2.49      0.0403
 rep#var|season |  14053.6693    36  390.379702
  --------------+----------------------------------------------------
                |
       Residual |     5075.75    28  181.276786
  --------------+----------------------------------------------------
   Total  |  138768.567    83  1671.91045
```

The results indicate that there were differences between the varieties as well as there being a season-by-variety interaction. To see these differences and the interaction, enter the following command for the results:

```
table var season, contents(mean wt)
```

```
-----------------------------------
               | Season: 1-spring,
               |       2-fall
       Variety |   Spring      Fall
---------------+-------------------
Orange Bulldog | 100.3375     126.6
      Longface |    .8125         0
   Sppktacular |  27.8375     6.375
        Spirit |      1.3       8.7
   Appalachian |      3.5      39.3
       Phantom |    .4625     1.925
     Trickster |     3.85     1.375
-----------------------------------
```

Although we did not do any more analysis other than to calculate the means, I think it is evident that Orange Bulldog has yielded significantly more than the other varieties. In addition, the interaction effect is rather modest.

Another common type of factorial is an evaluation over years. In this case, the years are not considered a fixed effect as seasons are, but rather a random effect, as every year can have different environmental effects. To show this analysis, open the dataset Plum Trial Years.dta and enter the following command:

```
anova yield year/rep|year var/year#var/rep#var|year
```

This results in the following output:

```
                Number of obs =       144     R-squared     =  1.0000
                Root MSE      =         0     Adj R-squared =

       Source |  Partial SS    df       MS           F     Prob > F
   -----------+----------------------------------------------------
        Model |  10961.7903   143  76.6558764
              |
         year |   719.53614     2   359.76807       4.10     0.0381
     rep|year |  1317.46597    15  87.8310646
   -----------+----------------------------------------------------
          var |  656.593453     7  93.7990647       1.10     0.4132
     year#var |  1191.44855    14   85.103468
```

```
-------+----------------------------------------------------------
    year#var |  1191.44855     14   85.103468       1.26     0.2432
rep#var|year |  7076.74621    105   67.397583
 --------+----------------------------------------------------------
             |
    Residual |           0      0
 --------+----------------------------------------------------------
       Total |  10961.7903    143  76.6558764
```

The results of this analysis indicate there are no treatment effects and the treatment by year also is not significant. The important idea to recognize in analyzing data over seasons or years is that seasons are considered fixed effects and years are considered random effects. This is evident in the selection of the denominator for calculating the F value. In the former case, the pooled error will have a larger degrees of freedom, which will result in a smaller denominator value. This means that finding differences between treatments will generally occur more often. In the latter case, the denominator degrees of freedom for treatment effects is the year-by-treatment interaction, which will have a smaller degrees of freedom and result in a larger denominator value for the F calculation and less chance of finding differences.

### Three-Factor Design

A three-factor design includes an additional factor for analysis and allows for both pairwise and three-way interactions to be analyzed. In theory, any number of factors can be analyzed in this fashion; however, in practical terms, such experiments become difficult to execute because of the size of the experiment and costs involved.

For example, a three-factor experiment might include variety, fertility program, and planting date. Load the file Three factor.dta, which is a dataset of onion yield with four varieties, five fertility levels, and three planting dates. This 4 × 5 × 3 factorial experiment has 60 treatments with four replications. It is evident how quickly such experiments can become quite large and unwieldy. Enter the following command:

```
anova wtlbs rep fertility sowingdate variety
fertility#sowingdate fertility#variety
sowingdate#variety fertility#sowingdate#variety
```

This results in the following output:

```
                         Number of obs =       240     R-squared     =  0.8428
                         Root MSE      = 17.2671      Adj R-squared =  0.7877

Source          |  Partial SS     df        MS           F       Prob > F
----------------+--------------------------------------------------------
          Model |  282950.468     62    4563.71722     15.31      0.0000
                |
            rep |  17324.2156      3    5774.73854     19.37      0.0000
      fertility |  131797.332      4   32949.3331     110.51      0.0000
     sowingdate |   91867.921      2   45933.9605     154.06      0.0000
        variety |  4741.19086      3    1580.39695      5.30      0.0016
fertility#
sowingdate      |  18378.6525      8    2297.33156      7.71      0.0000
fertility#
variety         |  949.869152     12      79.1557626    0.27      0.9935
sowingdate#
variety         |   14911.549      6    2485.25816      8.34      0.0000
fertility#
sowingdate#     |
   variety      |  2979.73751     24     124.155729     0.42      0.9931
                |
       Residual |  52773.3384    177     298.154454
----------------+--------------------------------------------------------
          Total |   335723.806   239    1404.70212
```

Looking at the results, all three factors (fertility, sowing date, and variety) are significant; however, there also are significant interactions for fertility × sowing date and variety × sowing date. Therefore, these interactions should be examined more closely. To begin with, it may be helpful to examine both the fertility and variety means over the different sowing dates. To do this, enter the commands

```
table fertility sowingdate, contents(mean wtlbs)
table variety sowingdate, contents(mean wtlbs)
```

This results in the following output tables:

```
----------------------------------------------------------
Fertility |
:0-200    |   Sowing Date: 1-10/5/01, 2-10/15/01,
lbs/acre  |                3-10/29/01
nitrogen  |  5 Oct. 2001  15 Oct. 2001  29 Oct. 2001
----------+-----------------------------------------------
        0 |      9.68125          8.39        3.4625
       50 |      72.6094        62.925       25.9562
      100 |      82.2906        71.315       29.4312
      150 |      96.8125          83.9       34.6125
      200 |      91.9719        79.705        32.875
----------------------------------------------------------
```

```
------------------------------------------------------------------
Variety:1-Nir |
vana,         |
2-Pegasus,    |
3-PS 7092,    |   Sowing Date: 1-10/5/01, 2-10/15/01,
4-Sweet       |                3-10/29/01
Vidalia       |  5 Oct. 2001  15 Oct. 2001  29 Oct. 2001
--------------+---------------------------------------------------
      Nirvana |     59.8965        53.801          27.74
      Pegasus |     71.7955        61.466          44.16
      PS 7092 |      72.343        58.546         19.385
Sweet Vidalia |     78.6575        71.175          9.785
------------------------------------------------------------------
```

In the first table, the results appear similar over the three sowing dates. The differences appear to be largely the magnitude of the yield with the third sowing date, 29 Oct. 2001, having much lower yields overall compared to the 5 or 15 Oct. 2001 sowing date. An examination of the second table of variety by sowing date indicates a difference in variety ranking for each of the separate sowing dates. A further examination of the separate analyses of variance (data not shown) for each of these sowing dates indicated that only on the 29 Oct. 2001 sowing date was there significant differences in the variety means.

These results also can be seen with the following commands: **margins** fertility#sowingdate, **marginsplot** and **margins** variety#sowingdate, **marginsplot**. The order of the fertility and sowingdate or variety and sowingdate will affect the **marginsplot** graph. Figure 6.7 shows the **margins** variety#sowingdate followed by the **marginsplot** command in the first graph. The second graph had **margins** sowingdate#variety entered followed by the **marginsplot** command.

### Split-Split Plot Design

The split-split plot design is an example of a three-factor experiment where the layout of the experiment is such that the factors occur as a main-plot effect with a subplot effect and finally a sub-subplot effect (Table 6.4). Generally the precision with which the factors can be analyzed increases from the main-plot effect to the sub-subplot effect. For this reason, if possible, the factor of most importance should be assigned to the sub-subplot, which has the greatest precision. This can

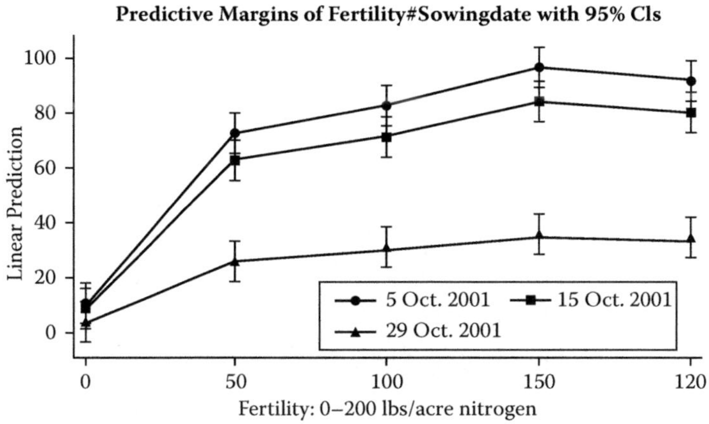


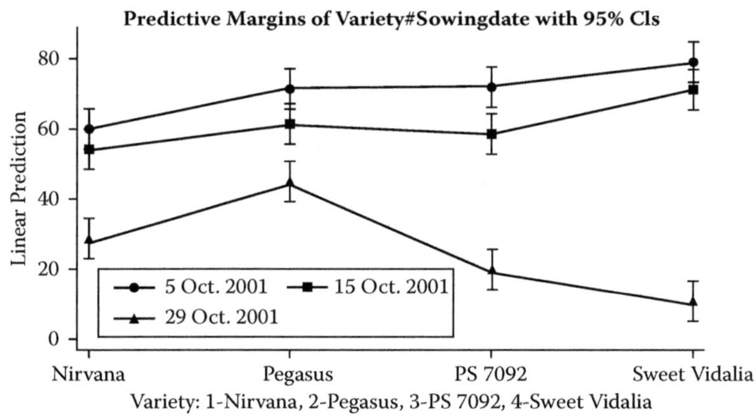


**Figure 6.7** The first graph using the `marginsplot` command after `margins fertility#sowingdate` and the second graph using the `marginsplot` command after the `margins` variety#sowingdate.

be seen in this example where the error degrees of freedom increases from the main plot, to the subplot, and finally the sub-subplot. Because the error sum of squares are divided by the error degrees of freedom and this is used as the denominator in an F-test, it is easy to see how the precision and ability to detect differences would increase. The arrangement of this design assigns first the main-plot effects randomly to the largest unit within the experiment. The subplot factors are then assigned randomly within the main plots and finally the sub-subplot factors are randomly assigned to within the subplots.

Figure 6.8 shows the layout of a split-split plot design with three replications arranged with five main-plot nitrogen fertility treatments,

**Table 6.4** Source of variation and degrees of freedom for a split-split plot design

| SOURCE OF VARIATION | DEGREES OF FREEDOM (DF) | RESULTS OF DF |
|---|---|---|
| Replication `(rep)` | $r-1$ | $3-1=2$ |
| Main plot `(nitro)` | $a-1$ | $5-1=4$ |
| Main-plot error `(rep#nitro)` | $(r-1)(a-1)$ | $(3-1)(5-1)=8$ |
| Subplot factor `(manage)` | $b-1$ | $3-1=2$ |
| Nitrogen × management interaction `(nitro#manage)` | $(r-1)(b-1)$ | $(5-1)(3-1)=8$ |
| Subplot error `nitro(rep - 1) (manage - 1)` | $a(r-1)(b-1)$ | $5(3-1)(3-1)=20$ |
| Sub-subplot factor `(var)` | $(c-1)$ | $(3-1)=2$ |
| Nitrogen × variety `(nitro#var)` | $(a-1)(c-1)$ | $(5-1)(3-1)=8$ |
| Management × variety `(mange#var)` | $(b-1)(c-1)$ | $(3-1)(3-1)=4$ |
| Nitrogen × management × variety `(nitro#mange#var)` | $(a-1)(b-1)(c-1)$ | $(5-1)(3-1)(3-1)=16$ |
| Sub-subplot error `nitro#manage(rep - 1) (var - 1)` | $ab(r-1)(c-1)$ | $5*3(3-1)(3-1)=60$ |

*Note:* Arrows indicate the ratio of mean squares for calculating F values.

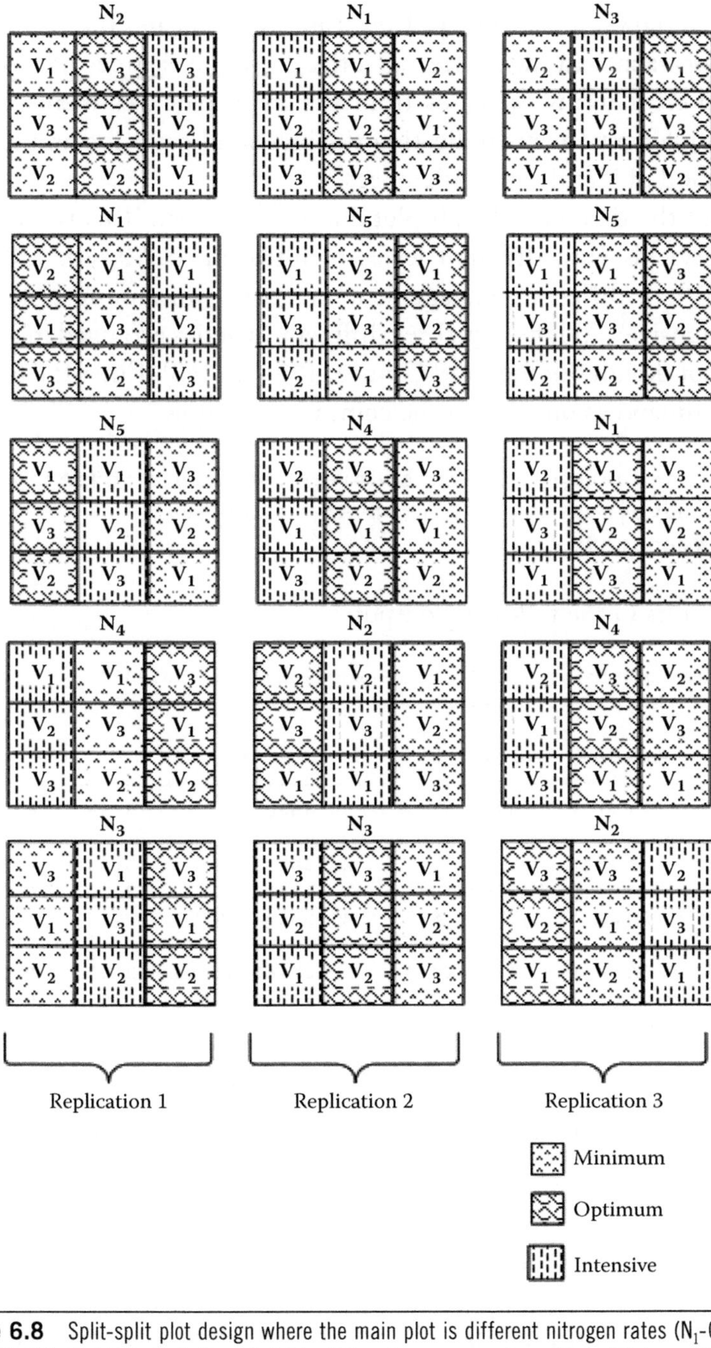


**Figure 6.8** Split-split plot design where the main plot is different nitrogen rates ($N_1$-0, $N_2$-50, $N_3$-80, $N_4$-110, $N_5$-140 kg/ha), the subplot is different management practices (Minimum, Optimum, and Intensive), and the sub-subplot effect is three different varieties ($V_1$, $V_2$, $V_3$).

three subplot management practices, and three sub-subplot varieties (Gomez and Gomez, 1984, p. 139). Table 6.4 indicates the error terms used for each treatment effect. Notice how the degrees of freedom for the error terms increases from the main-plot treatments to the sub-subplot treatments indicating the increased precision and greater likelihood of identifying differences.

Load the dataset Splitsplitplot.dta into memory. This is a dataset of rice yields with different levels of nitrogen (0, 50, 80, 110, and 140 kg/ha), different management practices (minimum, optimum, and intensive), and, finally, three different varieties arranged in a split-split plot design (Gomez and Gomez, 1984, p. 143). Enter the following command as one line in the command window:

```
anova yield rep nitro/nitro#rep manage nitro#manage/
rep#manage|nitro var nitro#var manage#var
nitro#manage#var
```

This results in the following output:

```
                  Number of obs =       135     R-squared     =  0.9204
                  Root MSE      = 703.947     Adj R-squared =  0.8222

                  Source |  Partial SS    df       MS          F     Prob > F
          ---------------+----------------------------------------------------
                   Model |   343808249    74  4646057.42     9.38    0.0000
                         |
                     rep |  731994.504     2  365997.252     0.66    0.5439
                   nitro |  61640821.8     4  15410205.5    27.70    0.0001
               nitro#rep |  4451350.68     8  556418.835
          ---------------+----------------------------------------------------
                  manage |    42936107     2  21468053.5    82.00    0.0000
            nitro#manage |  1102973.26     8  137871.657     0.53    0.8226
        rep#manage|nitro  |   5236334.81    20   261816.741
          ---------------+----------------------------------------------------
                     var |   206013160     2   103006580   207.87    0.0000
               nitro#var |  14144506.3     8  1768063.29     3.57    0.0019
              manage#var |  3851769.19     4  962942.296     1.94    0.1149
        nitro#manage#var |  3699232.07    16  231202.005     0.47    0.9538
                         |
                Residual |  29732489.3    60  495541.489
          ---------------+----------------------------------------------------
                   Total |   373540739   134  2787617.45
```

The results indicate that all three factors (fertilizer, management, and variety) affected rice yield. In addition, there was a fertilizer by variety interaction. There was no interaction effect for nitrogen

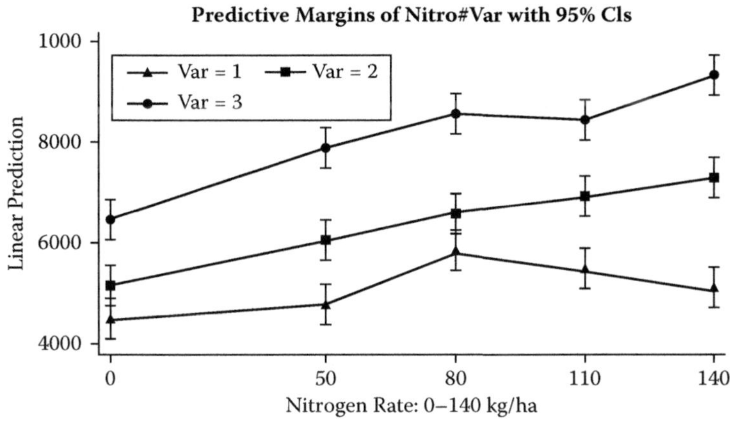


**Figure 6.9** Graphic output from the `marginsplot` command following `margins nitro#var` command.

fertilizer with management, management with variety, or a three-way interaction of nitrogen fertilizer, management, and variety.

The fertilizer by variety interaction can be further examined with the following commands:

```
margins nitro#var
marginsplot
```

Both varieties 1 and 2 (var) increase in yield with increasing fertilizer application. Variety 3, on the other hand, increases yield up to 80 kg/ha at which point the yield decreases (Figure 6.9).

### Covariance Analysis

Covariance analysis is a type of ANOVA that combines categorical and continuous factors to more accurately predict the effects of the categorical independent variable. This analysis of covariance that includes both categorical and continuous factors has as an underlying premise that there is a known relationship between the covariate and the independent variable. There are several conditions that should be met before using covariance analysis. The first is that the covariate is fixed and the covariate will not affect the treatments. Second, the regression of the covariate on the dependent variable is linear and independent of the treatments. Finally, the residuals are normally and independently distributed.

Analysis of covariance can be used in a number of different situations. It can be used to estimate missing data, to control experimental error, to adjust treatment means, and as an aid to experimental interpretation.

In controlling experimental error, covariance analysis introduces the covariate, which is considered to have an effect on the dependent variable. This effect, when removed, will generally lower the mean square error or residual. Load the dataset Covariance.dta. This is a dataset of a lima bean variety trial with 11 varieties arranged as an RCBD with four replications (Steel and Torrie, 1980, p. 412). Enter the command

```
anova ascorbic var rep
```

then enter the command

```
anova ascorbic var rep c.cov
```

This results in the two anova tables:

```
                 Number of obs =        55     R-squared     =  0.9040
                 Root MSE      = 12.1935       Adj R-squared =  0.8704

   Source |  Partial SS    df         MS             F     Prob > F
----------+----------------------------------------------------
    Model |  55987.1188    14   3999.07991       26.90     0.0000
          |
      var |  51018.1786    10   5101.81786       34.31     0.0000
      rep |  4968.94012     4   1242.23503        8.35     0.0001
          |
 Residual |  5947.30397    40   148.682599
----------+----------------------------------------------------
    Total |  61934.4227    54   1146.93375


                 Number of obs =        55     R-squared     =  0.9644
                 Root MSE      = 7.51657       Adj R-squared =  0.9507

   Source |  Partial SS    df         MS             F     Prob > F
---------+-----------------------------------------------------
    Model |  59730.9684    15   3982.06456       70.48     0.0000
          |
      var |  7457.62247    10   745.762247       13.20     0.0000
      rep |  756.392711     4   189.098178        3.35     0.0190
      cov |  3743.84965     1   3743.84965       66.26     0.0000
          |
 Residual |  2203.45433    39   56.4988289
---------+-----------------------------------------------------
    Total |  61934.4227    54   1146.93375
```

In the first instance, an RCBD is estimated without a covariate. In the second instance, the covariate cov is introduced into the model. In a lima bean variety trial, it is difficult to harvest the crop where each entry is at the same maturity and it is known that the more mature the beans, the lower the ascorbic acid content. This covariate is then percent dry matter content, which is an indicator of maturity. The **c.** in front of this variable is telling the program to treat this variable as a continuous variable and calculate it as a regression (see Chapter 10, Correlation and Regression). In the second model, the residual mean square is lower (148.682599 versus 56.4988289) and both the $R^2$ and adjusted $R^2$ are higher. This indicates that the second model is more precise in its estimate of the treatment effect.

With analysis of covariance, it is customary to present adjusted or marginal means rather than the simple arithmetic means. The marginal means take into account the effect of the covariate and can have a significant impact on the interpretation of results. The adjustment of the means is calculated as follows:

$$\hat{Y}_{i\cdot} = \bar{Y}_{i\cdot} - b_{YX}(\bar{X}_{i\cdot} - \bar{X}_{\cdot\cdot})$$

The $\hat{Y}_{i\cdot}$ is the adjusted treatment mean. The $\bar{Y}_{i\cdot}$ represents the observed mean and the $b_{YX}$ is the error regression coefficient. The $\bar{X}_{i\cdot}$ and $\bar{X}_{\cdot\cdot}$ represent the observed covariance mean and the overall mean for all the covariance entries, respectively. To calculate the adjusted mean for the first entry (lima bean variety), which has an observed mean of 88.1, enter the following commands:

```
tabstat ascorbic, statistics(mean) by(var)
columns(variables)
tabstat cov, statistics(mean) by(var) columns(variables)
tabstat cov, statistics(mean) columns(variables)
anova ascorbic var rep c.cov
matrix list e(b)
```

The first two **tabstat** commands calculate the means for the ascorbic (ascorbic acid content) and cov (dry weight percentage) variables by var (varieties). This gives us the mean for the first variety (i.e., 88.1) and the mean for the corresponding covariate (i.e., 35.42). The third **tabstat** command calculates the overall mean for cov (i.e.,

33.98727). The fourth command calculates the analysis of covariance as shown above and the last command, **matrix list** e(b), which must be entered immediately after the **anova** estimation command, shows the regression matrix for the analysis of covariance. The error regression coefficient is shown as part of this matrix below cov with a value of –3.1320175. Entering the values in the adjusted mean equation results in the adjusted mean for the first variety of 92.58733.

$$92.58733 = 88.1 - (-3.1320175)(35.42 - 33.98727)$$

This is a long way to go and rather tedious to calculate these means. Fortunately, Stata has a command that calculates the adjusted or marginal means. Enter the following command:

```
margins var
```

This results in the following output:

```
Predictive margins                Number of obs    =          55

Expression    : Linear prediction, predict()

------------------------------------------------------------------
       |     Delta-method
       | Margin  Std. Err.     z    P>|z|  [95% Conf. Interval]
-------+----------------------------------------------------------
   var |
    1  |92.58733 3.406412  27.18  0.000 85.91088      99.26377
    2  |79.11642 6.275393  12.61  0.000 66.81688      91.41597
    3  |78.10311 3.508606  22.26  0.000 71.22637      84.97985
    4  |84.53012 3.950303  21.40  0.000 76.78767      92.27257
    5  |95.98305 4.967951  19.32  0.000 86.24605      105.7201
    6  |97.50684 3.588881  27.17  0.000 90.47277      104.5409
    7  |99.97868 3.379806  29.58  0.000 93.35438       106.603
    8  |72.04475 3.364251  21.41  0.000 65.45094      78.63856
    9  |81.14672 3.390371  23.93  0.000 74.50171      87.79172
   10  |122.7838 5.140306  23.89  0.000 112.709       132.8587
   11  |74.31913 3.365915  22.08  0.000 67.72206      80.91621
------------------------------------------------------------------
```

The Margin column lists the marginal or adjusted means. You will note that the rank ordering of these means will differ considerably from the observed means. Previously when we used the **margins** command, the means were the arithmetic means because there was no covariate involved.

There are a number of examples of using a covariate to eliminate some factor that is affecting the outcome and to more precisely calculate the residual. One is stand count, which is used as a common covariate in many experiments. This is useful where the plant stand is not complete because of poor germination or adverse effects after transplanting.

Other examples of covariate use involve the initial weight of experimental animals where the gain in weight is the dependent variable and this gain in weight may be affected by the animal's initial weight.

Field position also may be used as a covariate. For example, an experiment may have been planted near the edge of a field where there is a distinct edge effect. Perhaps plants along the field's edge may be robbing nutrients and water from your experimental plants, which, in turn, could affect your results. In this case, using the reciprocal of the distance from the field's edge would be an appropriate covariate. By using such a covariate, the farther from the field's edge, the lower the effect. This type of effect would generally be taken care of by blocking, but in some cases such effects may not be completely evident at the start of an experiment.

Another example is soil heterogeneity and its effect on treatment effects. Blocking, as in an RCBD, can have a significant impact in reducing effects due to plot location in a field. But sometimes soils can be heterogeneous in such a way that blocking cannot easily deal with the problem. A uniformity trial prior to experimental work can identify such soil heterogeneity and these data can be used as a covariate.

Analysis of covariance can be used to estimate missing data as well. To estimate missing data and complete the analysis, first set the missing data point to 0, then set up a covariate that has values of 0 for all data points except the one with the missing value, which should be set to 1. Then conduct the analysis of covariance. To see this, we will use the Covmissing.dta dataset. This is a dataset of ascorbic acid content in turnip greens with three treatments of postharvest handling (Steel and Torrie, 1980, p. 427). Replace the missing data point with a 0. Then create a new covariate (i.e., X) with values of 0 for all entries except for the data point that is missing, which will have a value of 1. This is often referred to as a dummy variable. The following commands will accomplish this:

```
replace ascorbic = 0 if ascorbic ==.
generate x = 0
replace x = 1 if ascorbic == 0
```

After adding the changes to the dataset, enter the following command:

```
anova ascorbic trt rep c.x
```

This results in the following output:

```
                  Number of obs =        15     R-squared     =  0.9569
                  Root MSE      = 76.2643       Adj R-squared =  0.9139

         Source |  Partial SS    df        MS            F     Prob > F
    ------+---------------------------------------------------------
          Model |  904582.008     7  129226.001      22.22     0.0003
                |
            trt |  20246.9417     2  10123.4708       1.74     0.2435
            rep |   49805.025     4  12451.2563       2.14     0.1785
              X |  341226.675     1  341226.675      58.67     0.0001
                |
       Residual |   40713.725     7  5816.24643
     -----------+---------------------------------------------------------
          Total |  945295.733    14  67521.1238
```

This results in an unbiased partial sum of squares for the treatments (i.e., trt) and can be used to estimate a value for the missing data point. After entering the estimation command, enter

```
matrix list e(b)
```

The value for –x in this matrix is an unbiased estimate for the missing value, which is –(–799.875) or 799.875. This value can then be substituted in the dataset for the missing value and the ANOVA run again as

```
anova ascorbic trt rep
```

This results in the following output:

```
                  Number of obs =        15     R-squared     =  0.6721
                  Root MSE      = 71.3387       Adj R-squared =  0.4262

         Source |  Partial SS    df        MS            F     Prob > F
       --------+---------------------------------------------------------
          Model |       83462     6  13910.3333       2.73     0.0948
                |
```

```
     trt |  25292.9333     2  12646.4667        2.48     0.1447
     rep |  58169.0667     4  14542.2667        2.86     0.0964
         |
Residual |  40713.7333     8  5089.21667
---------+----------------------------------------------------
   Total |  124175.733    14  8869.69524
```

Because there was a missing value, the degrees of freedom for the residual or error is lowered from 8 to 7, and a new residual mean square calculated. Thus, the mean square error is now 5,816.2476 (40,713.7333/7) and the F value for treatments (trt) is 2.1743343 (12,646.4667/5,816.2476). To calculated and display the new probability after this adjustment, the **Fden**(n1,n2,f) density function can be used. The n1 is the degrees of freedom for the numerator (i.e., 2) and the n2 is the degrees of freedom for the denominator (i.e., 7). The f is the calculated F (i.e., 2.1743343) value.

```
display Fden (2,7, 2.1743343)
```

which results in

```
.11368131
```

This procedure can be used for more than one missing value. A new dummy variable would be created for each additional missing value and the process repeated, including reducing the error degrees of freedom by one for each missing value. Obviously, there is a limit to the number of missing values you should replace. In a planned experiment such as this, you would expect very few missing values, but it does happen. The decision to continue an analysis with multiple missing values is a judgment call. Remember, statistics is a tool to help you understand your data, not a crutch to hold up an experiment with problems.

# 7
# Programming Stata

Stata, for the casual user, offers a nicely implemented GUI (graphical user interface), which makes it easy to use, but its real strength lies in its expandability with user-written routines. It may be surprising, but a large part of Stata is actually written and implemented with its own built-in language. These programs can actually be viewed by the user wishing to see how a function is implemented or to learn more about programming. These files are stored in the Stata folder in the Applications folder (locations may be different based on operating systems). When your program is updated, the updates often contain many of these programs. In addition to these official updates, it is possible to download and use user-written programs that expand Stata's capabilities. You may be interested in which commands are built into the Stata program and which are written as Stata commands. Stata has a command to do just that. Enter the following:

```
which anova
```

This results in the following output:

```
/Applications/Stata/ado/base/a/anova.ado
*! version 2.1.0 07jun2011
```

This output indicates the pathname to where the file is located, the file's internal version number, and the date of its latest change. The pathname above is how it will appear on a Macintosh and will appear slightly different on a Windows or Unix computer. If, however, you entered

```
which generate
```

the output would be

```
built-in command: generate
```

Stata uses the convention of adding .do or .ado as extensions to program files. These files are actually just text files that Stata interprets as executable programming code. The .do extension refers to a do-file as it is called because it does something. These files must first be loaded into computer memory before they can be used by the user. It differs from an ado file in that an ado file is loaded into memory and executed in one step. That is, a command that is implemented as an ado file when typed into the Command window in the correct format will automatically load and run. This makes the ado files seamless to the user, acting as if they are part of the Stata program. If the do-file has been saved it also can be invoked by using the **do** command followed by the file name. Stata will then look in specific directories and the working directory for the file and execute the commands in the so-named do-file. If it cannot find the file, it will return a `file not found` error.

One of the easiest uses of the programming capabilities of Stata is to use the Do-File Editor to handle a series of commands. A number of different commands can be executed at one time in a single file. This can be particularly helpful when similar analysis is done on several different variables. To demonstrate how this works, load the data file Large Onion Dataset 2001-02.dta. This dataset includes several variables from an onion variety trial. These variables include various yield components (e.g., yield, drywts, jumbo, and mediums), as well as onion quality parameters (e.g., pungency, sugar, and doubles). As a starting point to analyzing these data, you might want to look at an analysis of variance and the means for all of these variables. This could be done by entering the **anova** and **tabstat** commands in the Command window one after the other or select the commands from the menus for all the variables. This might become tedious and can be accomplished more quickly in the Do-File Editor. Open a new do-file window by selecting from the Do-File Editor under the Windows menu. Then enter the following commands:

```
anova yield variety rep
tabstat yield, by(variety)
```

These commands then can be copied and pasted into the same window. You will want to paste these commands in the window six times.

Then change yield in these pasted commands to pungency, sugar, drywts, jumbo, mediums, and doubles. You may wish to have a log file started before executing these commands, which can be part of this do-file. In addition, you can include the file name in the do-file so that the do-file will open the dataset, start a log, complete the analysis, and close the log. To see the complete do-file, open Large dataset.do. Note for this to work properly you will have to change your working directory to where the Large Onion Dataset 2001-02.dta dataset is located. To change the working directory select the Change Working Directory… item under the File menu. This also can be accomplished with the **cd** command, but will have a slightly different pathname based on the operating system used (i.e., Macintosh, Windows, or Unix). Then select the icon Do in the top of the Do-File Editor window. This file will quickly analyze all the variables and list the treatment means. Also included in this do-file is the command **set more off**. This turns off the **more** function in the Results window so the results do not pause for each window of data, but rather runs quickly through all of the analyses. Because a log file was created, this can be opened from the Log menu under File. This file will have a .smcl extension.

Stata programs can be much more than just files of commands. They can be written to act like other ado files, such as official Stata ado files. This means they can include Help files and GUI elements. They can be shared with other users or downloaded and installed on your computer. They can be written just for the specific problem at hand or they can be written to be reused for similar problems. They can be complex or simple.

If you are familiar with computer programming, many of the conventions, structure, and program control will appear familiar. There are, however, some differences that can initially appear confusing. On the other hand, for the neophyte, it is relatively easy and straightforward to write and use programs and their usefulness will quickly become evident.

Some of the issues that can be confusing in Stata programming deal with the vocabulary used in discussing the language. For example, Stata uses the term *macro* to indicate what in other languages is called a *variable*. Remember, however, that Stata is primarily a statistics program, not a programming language, and, consequently, the term *variable* is reserved for the columns of data in a dataset. This also can be confusing to Microsoft Office® users where macros are programs implemented within Microsoft Office. Microsoft has its Visual Basic for Applications (VBA), which is an implementation of Visual Basic that is used within the Microsoft Office environment where these programs are called macros.

Stata programs can actually be entered interactively, but rarely are. Here is an example:

```
. program quote
  1. display "Now is the time for all good men to come
to the aid of their country."
  2. end
```

Once this program (quote) has been defined, all you have to do is type the word *quote* and the program executes displaying the quotation. Usually programs are entered in a do-file so that the program can be used over and over again. In addition, as programs become more complex, you will need to make corrections or debug them before they run correctly. Once a program has been defined in Stata's memory it cannot be redefined. Thus, if you entered `program quote` after having defined it by our example above, Stata will return an error message that the program is already defined. To prevent this from happening, the first line in the do-file should be

```
capture program drop quote
```

The **capture** command executes any subsequent command and suppresses any error codes. The **drop** command drops the quote program before redefining it. If the program was not in memory and we just had the **drop** command, it would return an error code, so that's why **capture** is included.

Before beginning a program, let's take a closer look at the Do-File Editor and how Stata handles such files. This editor window can be opened either from the Main window by selecting the icon that looks like a notepad and pencil or from the Windows' menu. Saved do-files can be opened from the File menu or selecting the Folder icon in the Main window.

The Do-File Editor has several icons across the top of the window, which will be somewhat different depending on the operating system

(Macintosh, Windows, or Unix). On the Macintosh, in order from the left side are icons used for opening a do-file, saving your do-file, printing the do-file, searching the open do-file, and showing paragraph marks. On the right top of the Do-File Editor are the Run and Do icons. The Run icon executes the program without echoing results to the Results window, while Do does echo the results to the Results window.

Both Windows and Unix computers' Do-File Editor will have a different appearance, but will have the same overall functionality. On Windows computers, because menus are integrated into the window, much of the functionality is found under these menus. There also are several icons across the top of the window for saving, opening, and creating new do-files. Also available are icons for cut, copy, and paste as well as undo and redo. There are also icons for searching the current file and for executing the do-file (Figure 7.1).

A nice new feature of the Do-File Editor is colors for different elements in a program. Different colors can be chosen for commands, functions, comments, strings, variable types, macros, and numbers. These features can be accessed from the Preferences item under the Stata menu. On a Unix or Windows platform, click Edit in the Do-File Editor and then select Preferences. This can make reading and debugging programs a lot easier and can make your programs more readable by others.

Line numbers also can be added to the editor from the preferences and when a line number is selected a bookmark is added. You can use this feature to quickly navigate through a program.

Other features available in the Preferences for the Do-File Editor include auto-tabbing, opening new do-files as either new tabs or new windows on a Macintosh. New windows are not available on a Windows platform, but the same functionality of viewing two files simultaneously is available by dragging the tab for a file into the viewing area.

Let's begin to use Stata's programming capabilities with a simple program that expands the usefulness of Stata. Figure 7.1 shows the Do-File Editor with the complete program. The complete program is available as anovalsd.do, which can be loaded to see the different colors associated with different elements of the program. Although the program is available on disk, let's go ahead and start a new Do-File Editor screen. This program will be called *anovalsd*,

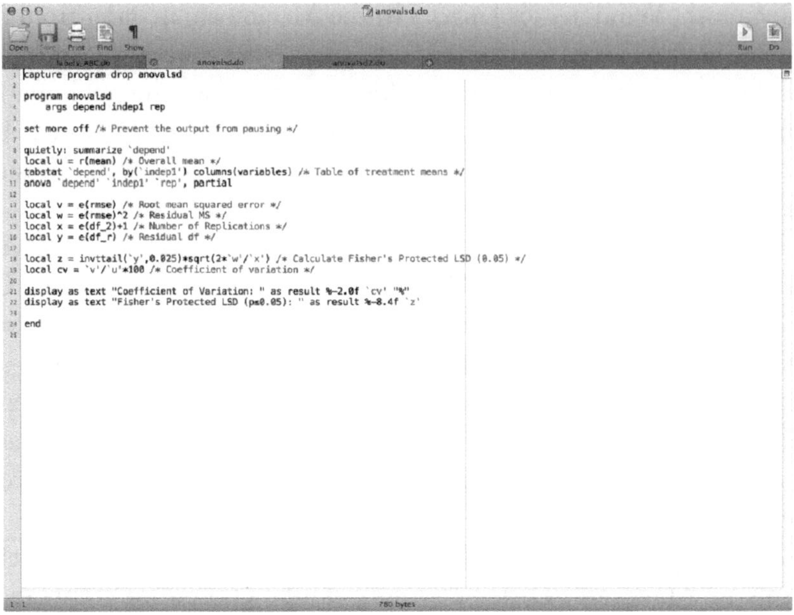


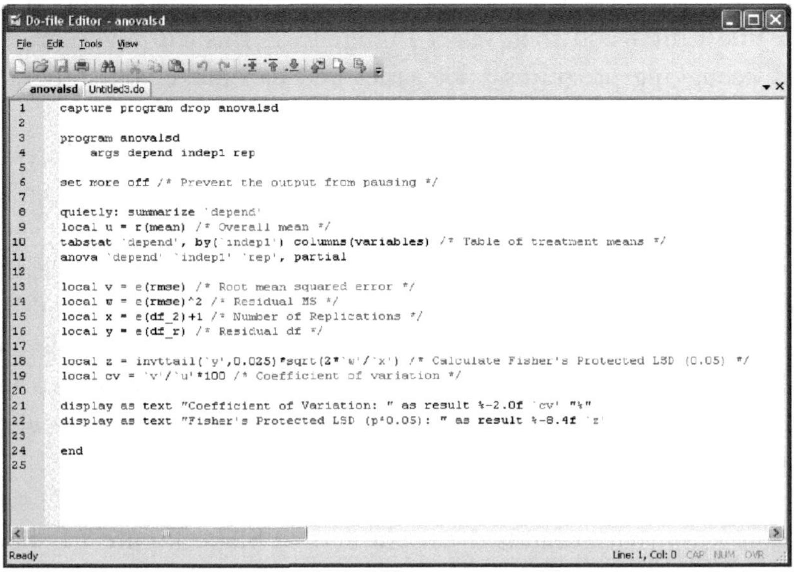


**Figure 7.1** Do-File Editor with the anovalsd.do file visible on the Macintosh (above) and Windows computer (below).

so the first step in writing this program is to enter **capture program drop** anovalsd. Remember this line drops the program anovalsd from memory and ignores any error code that may occur if the program is not in memory. The next line to enter is **program** anovalsd, which defines the program. The next line **args** tells the program that subsequent items are arguments, which will be used in the program. When this program is run, it requires three arguments, which are passed to the program. To use the program after it has been run, you would enter **anovalsd** depend indep1 rep where the three arguments would be the variable names from the dataset in memory representing the dependent, independent, and replication variables.

The next line **set more off**, as mentioned previously, turns off the pause feature in the Results window. Usually output is paused with every screen requiring the user to hit a key to see the next screen. By turning off this feature, the results are displayed all at once without pausing.

**quietly: summarize** \`depend' is one of the great features of Stata programming. Executing a command **quietly** means no output is to be generated. Instead the command **summarize** is executed and values calculated by this command are stored in memory. These saved results then can be used by subsequent commands. To demonstrate this, load the dataset Onion varieties programming.dta and enter the following command:

```
summarize Yield
```

This results in the following output:

```
    Variable |       Obs        Mean    Std. Dev.       Min        Max
-------------+--------------------------------------------------------
       Yield |        20       102.4    10.43259       82.5      122.5
```

Now enter the command

```
return list
```

This results in the following output:

```
scalars:
                  r(N)  =  20
              r(sum_w)  =  20
               r(mean)  =  102.4
                r(Var)  =  108.83894698294
                 r(sd)  =  10.43259061704906
                r(min)  =  82.5
                r(max)  =  122.5
                r(sum)  =  2048
```

These are referred to as scalars because they represent specific values and these values can be used in subsequent operations by referring to their label (`r(N), r(sum _ w),` etc.). These scalars are only held temporarily in memory. If another summarize command with a different variable were entered then the new scalars would be in memory.

The `r(mean)` value is then used by the next line in the program, **`local`** `u = r(mean)/* Overall mean */`. A local macro is one that can hold a value while this program is executing. As soon as the execution is complete, the local macro is dropped. Now the program has the `r(mean)` value (i.e., 102.4) stored in the local macro `u`. The remainder of this line `/* Overall mean */` is just a comment. Anything that appears between /* and */ is ignored by the program. As you write programs, it is a good idea to add comments explaining what is happening or identifying items. This will help you remember what you did when you come back to the program later or will help others see exactly what you have done. Comments also can be added using double slashes (//). The /* */ format is generally used for larger comments of several lines.

The next line,

```
tabstat `depend', by(`indep1') columns(variables)/*
Table of treatment means */
```

calculates a table of means for the first argument (`depend`) using the second argument (`indep1`) to group the means and places them in a column format. It is important at this point to explain the use of quotation marks. The open and closed quote marks tell Stata that the value of the macro should be used. If the quotes are not present, then Stata interprets it to mean just the word (depend, indep1, etc.). The open quote that we are using here is located next to the 1 in the upper left side of most keyboards and the close quote is located on the right of the keyboard between the semicolon and return keys. It is important to use these specific keys, unlike normal computer use where the open and close quotes used are the same key.

The **anova** `` `depend' `indep1' `rep', `` **partial** uses all three variables passed to the program to calculate an analysis of variance. The sequence of arguments when using this program is important. The first argument (`depend`) is considered the dependent variable, while the `indep1` and `rep` are independent variables. I do a lot of variety trial evaluations with vegetables, which are usually in an RCBD (randomized complete block design). So, for me, the `indep1` macro is for the variety list and the `rep` is the replication.

The next four lines create four macros (`v`, `w`, `x`, and `y`) that are values from the analysis of variance. Just as there were values saved after the **summarize** command, there are values saved after the **anova** command. The **summarize** command is an r-class command, while **anova** is an e-class command. To see the saved results after an analysis of variance, type

```
ereturn list
```

This returns several scalars as well as other information. The scalars we are interested in using include `e(rmse)`, which is the root mean square error from the most recent analysis of variance. The square of this value is the residual mean square. The `e(df _ 2)` is the degrees of freedom for the replications and adding 1 to this is then the number of replications, and, finally, the `e(df _ r)` is the residual degrees of freedom.

The next line calculates the Least Significant Difference (LSD) at the 5% level. The formula for this calculation is

$$LSD = t_{crit.}\sqrt{\frac{2MSE}{n}}$$

The $t_{crit.}$ value is the critical value of Student's t that can be found in tables at the back of statistics textbooks. Stata has a function to calculate this value:

```
invttail(n,p)
```

This function calculates the one-tailed Student's t, which requires the residual degrees of freedom (n) and the probability of interest (p). The residual degrees of freedom is from the previous ANOVA scalar (e(df _ r)). Because this function calculates the one-tailed Student's t and we are interested in the two-tailed value, the probability entered is half of the value we are interested in. Thus, for a 5% (0.05) level, we enter 0.025. If you are interested in seeing what this value is, enter

```
display invttail(12,0.025)
```

The value calculated is 2.1788128, which is the critical value for a two-tailed Student's t. The remainder of this line completes the equation shown above. The line with the comment left out is

```
local z = invttail(`y',0.025)*sqrt(2*`w'/`x')
```

The critical t value is multiplied (*) with the square root (**sqrt()**) of 2 multiplied by the mean square error (MSE) (`w') and divided by the number of replications (`x'). The previous ANOVA does not save the MSE as a scalar, but does save the square root of this value (Root MSE) in the scalar e(rmse). Squaring this value then gives us what we need (e(rmse)^2), which is 102.920659. By the way, * and ^ are arithmetic operators; to see a list of these, refer to Table 7.1. This information can be seen within Stata by typing **help operators** in the Command window.

**Table 7.1** Expression operators used in programming and various commands

| ARITHMETIC OPERATORS | | LOGICAL OPERATORS | | RELATIONAL OPERATORS | |
|---|---|---|---|---|---|
| SYMBOL | | SYMBOL | | SYMBOL | |
| + | addition | & | and | > | greater than |
| - | subtraction | \| | or | < | less than |
| * | multiplication | ! | not | >= | greater than or equal to |
| / | division | ~ | not | <= | less than or equal to |
| ^ | power | | | == | is equal to |
| - | negative | | | != | not equal |
| + | string concatenation | | | ~= | not equal |
| = | equals | | | | |

*Note:* A double equal sign (= =) is used for equality testing.
The order of evaluation (from first to last) of all operators is ! (or ~), ^, – (negation), /, *, – (subtraction), +, ! = (or ~ =), >, <, < =, > =, = =, &, and |.

The next line calculates the coefficient of variation (CV), which is usually the standard deviation divided by the mean, which is multiplied by 100 and reported as a percent. In this case, using the root MSE divided by the mean of the dependent variable and multiplied by 100 results in a similar value. The CV gives a unit independent value of the dispersion around the mean. In the context of an experiment, a smaller value is considered better and often indicates the overall experimental conditions or model fit. This value also can change based on the type of experiment or crop involved. Because this value is not reported in any particular unit (e.g., inches, pounds, lbs/acre, etc.), it can be used to compare the performance of different experiments with different units of measure. It is possible under some circumstances to have a CV that is over 100%. This can indicate a problem with the experiment or the nature of the collected data. In any event, values over 100% indicate that the means are of little value.

The next two lines display the output for this program. The **display** command displays what follows in the Results window and using **display as** is followed by a style that is determined by the color scheme set in the Preferences. The styles available are **text, result, error,** and **input**. The **text** style is used for identifying text. The **results** style is for the results of calculations. The **error** style is for displaying errors, and the **input** style is rarely used, but is generally reserved for user input. The uses of these styles are not set in stone, but rather are suggested uses for consistency across commands. To best see the effects of styles, change the preferences to the classic scheme and run this program.

The following lines within the program illustrate the use of these styles.

```
display as text "Coefficient of Variation:
" as result%-2.0f `cv' "%"
display as text "Fisher's Protected LSD (p≤0.05):
" as result%-8.4f `z'
```

The `%-2.0f` and `%-8.4f` are formatting directives. The `%` indicates what follows is for formatting. The `-` indicates it should be left justified. The number to the left of the decimal is the total output width and the number to the right of the decimal is the number of decimal places. The f indicates that it is a fixed format in terms of the number of decimal places. Stata has many detailed formatting directives including for dates and times.

Finally, the last line **end** ends the program. This simple program is one that I use routinely to calculate treatment means, CV, and the LSD value for variety trials.

One of the results lines lists its output as Fisher's Protected LSD ($p \leq 0.05$). This is not entirely correct. Actually, this line only calculated the LSD, which controls for comparisonwise error. In order for this LSD to be Fisher's Protected LSD, there needs to be an experimentwide level of significance; the treatment probability (Variety) should be below 5%. In the example dataset, this probability is not actually significant at the 5% level ($p = 0.3367$).

This then is an opportunity to improve the program. Several new lines of code will be added to this do-file to determine if, in fact, the experimentwide error rate is below the requisite 5% level. To begin, the following lines of code will be added after the CV is displayed. Actually, these lines can be added anywhere in the program as long as they occur after the **anova** command has been executed. These lines are

```
local r = e(df_1)
local s = e(df_r)
local t = e(F_1)
```

These lines save the scalars for the degrees of freedom for treatments (numerator) and the residual (denominator), as well as the calculated F value for the treatments, respectively. Their values are 4, 12, and 1.264906429141414. The next line added is

```
local a = Ftail(`r',`s',`t')
```

The **Ftail()** function calculates the probability, which, in this case, is associated with the variety differences. The probability calculated is 0.33669645. This is the same probability (with more decimal places) than is shown in the ANOVA table under the Prob > F heading for variety. This value is then used to compare with the level of significance chosen (i.e., 0.05). The next section of code to be added then makes a decision and displays results based on this comparison. The code is

```
if `a' > 0.05 {
     display as text "Fisher's Protected LSD is not
significant p = " as result%-8.4f `a'
          }
else {
     display as text "Fisher's Protected LSD (p≤0.05):
" as result%-8.4f `z'
          }
end
```

The **if** command evaluates an expression, in this case, `` `a' `` > 0.05, and, if it is true (nonzero in computer parlance), then the commands within the braces {} are executed. If the expression evaluates to false (zero), then the commands within the braces are skipped. In this case, with the **else** command, an alternative set of commands within the braces after the **else** command is executed.

To see this program work, load the dataset Onion varieties programming.dta. The anovalsd2.do file has these added lines and should be loaded and executed and then enter the following command:

```
anovalsd2 Yield Variety Replication
```

This results in the following output:

```
Summary statistics: mean
  by categories of: Variety (Variety Number)

 Variety |     Yield
---------+----------
       1 |   100.375
       2 |   104.825
       3 |     95.15
       4 |   110.525
       5 |   101.125
---------+----------
   Total |     102.4
--------------------

                Number of obs =       20     R-squared     =  0.4028
                Root MSE      =   10.145     Adj R-squared =  0.0544

   Source    |  Partial SS   df        MS            F      Prob > F
-------------+----------------------------------------------------
       Model |  832.892087    7  118.984584       1.16      0.3932
             |
     Variety |  520.740012    4  130.185003       1.26      0.3367
```

```
Replication |   312.152075    3  104.050692        1.01      0.4217
            |
   Residual |   1235.04791    12  102.920659
------------+--------------------------------------------------------
      Total |   2067.93999    19  108.838947
Coefficient of Variation: 10%
Fisher's Protected LSD is not significant p = 0.3367
```

There will be more information on programming in the next chapter. I thought it would be better to see how using programming can help solve real problems.

# 8
# Post Hoc Tests

## Planned Comparisons

Analysis of variance will answer the question: Are there significant differences between treatments and are there any interactions between factors when more than one factor is involved? Post hoc tests are performed after the ANOVA (analysis of variance) to answer the specific question about which treatments differ.

Oftentimes the experiment is such that logical comparisons between the treatments can be planned and evaluated. It is generally not recommended that comparisons be chosen based on the results. The comparisons of interest should be considered and planned in advance to avoid any bias. Load the dataset Rice Fertilizer Comparisons.dta and enter the command

```
anova yield trt rep
```

This dataset is of an experiment on rice yield with different fertilizers including ammonium sulfate ($NH_4SO_4$), green leaf (presumably some type of compost or organic matter), and a combination of both (Palaniswamy and Palaniswamy, 2006, p. 401). The ANOVA results are significant and there are comparisons of interest that were decided on in advance of the experiment. These include comparing the control to the fertilizer treatments, comparing $NH_4SO_4$ to the green leaf, and comparing the $NH_4SO_4$ and green leaf treatments to the combination of $NH_4SO_4$ and green leaf. Table 8.1 presents these results as orthogonal coefficients. Comparisons are considered orthogonal when the coefficients add to 0.

Immediately after the ANOVA, the **test** command can be used to make these specific single degree of freedom comparisons. The commands and output for the first comparison, control versus fertilization, are

**Table 8.1** Orthogonal coefficients for planned comparisons

| CONTRAST | CONTROL | $NH_4SO_3$ | GREEN LEAF | $NH_4SO_3$ + GREEN LEAF |
|---|---|---|---|---|
| Control versus fertilization | 3 | −1 | −1 | −1 |
| $NH_4SO_4$ versus green leaf | 0 | 1 | −1 | 0 |
| $NH_4SO_4$ and green leaf versus $NH_4SO_4$ + green leaf | 0 | 1 | 1 | −2 |

```
test 3*1.trt-2.trt-3.trt-4.trt=0

 ( 1)  3*1b.trt - 2.trt - 3.trt - 4.trt = 0

       F(  1,      9) =   82.57
            Prob > F =    0.0000
```

In addition to the **test** command, there is the **contrast** command that can be used. Enter the following and see the results:

```
contrast {trt 3 -1 -1 -1}

Contrasts of marginal linear predictions

Margins      : asbalanced

------------------------------------------------
             |         df           F        P>F
-------------+----------------------------------
         trt |          1       82.57     0.0000
             |
    Residual |          9
------------------------------------------------

--------------------------------------------------------------
             |   Contrast   Std. Err.     [95% Conf. Interval]
-------------+------------------------------------------------
         trt |
        (1)  |    -15.475      1.703     -19.32745   -11.62255
--------------------------------------------------------------
```

There is slightly more information shown with the **contrast** command, but the saved scalars from the **test** command are more useful.

Stata will evaluate any algebraic expression before calculating the **test** command. The additional comparisons are

```
test 2.trt - 3.trt = 0

 ( 1)   2.trt - 3.trt = 0
        F(  1,      9) =     1.49
             Prob > F =     0.2525

test 2.trt + 3.trt - 2*4.trt = 0

 ( 1)   2.trt + 3.trt - 2*4.trt = 0
        F(  1,      9) =     8.45
             Prob > F =     0.0174
```

The number preceding the variable reflects the level of the variable. This dataset has value labels associated with the levels of `trt` indicating what the treatments were. To see the actual levels (i.e., numbers) associated with `trt`, enter **`label list`** in the Command window. The above comparisons can be entered with different algebraic expressions and have the same result; however, this should be avoided to prevent mistakes in the contrasts. Entering them as they appear in Table 8.1 as orthogonal contrasts is a good habit to get into.

With a slightly more complex situation, load the dataset Corn Seed Treatments.dta, which is a dataset of corn stand counts from a greenhouse experiment of different fungicide seed treatments (Steele and Torrie, 1980, p. 206).* This dataset has eight treatments that include an untreated check, two types of mercuric fungicides, two types of nonmercuric fungicides from one company and three types of nonmercuric fungicides from a second company. The last three treatments include different formulations of the same material. The planned comparisons are shown in Table 8.2.

After loading the dataset, run an ANOVA (**`anova`** `stand trt rep`). The post hoc tests and the results of these planned comparisons are

```
test 7*1.trt - 2.trt - 3.trt - 4.trt - 5.trt - 6.trt - 7.trt
- 8.trt = 0

 ( 1)  7*1b.trt - 2.trt - 3.trt - 4.trt - 5.trt - 6.trt - 7.trt
- 8.trt = 0

       F(  1,     35) =     4.95
            Prob > F =     0.0327
```

* Mercury-based seed treatments are no longer allowed in agricultural production.

**Table 8.2** Planned comparisons for seed treatment fungicides

| CONTRAST | COMPARISONS | ORTHOGONAL COEFFICIENTS | | | | | | | |
|---|---|---|---|---|---|---|---|---|---|
| 1 versus 2–8 | Do the fungicides work compared to the check | 7 | –1 | –1 | –1 | –1 | –1 | –1 | –1 |
| 2–3 versus 4–8 | Compare the mercuric fungicides to the nonmercuric fungicides | 0 | 5 | 5 | –2 | –2 | –2 | –2 | –2 |
| 2 versus 3 | Compare the different mercuric fungicides | 0 | 1 | –1 | 0 | 0 | 0 | 0 | 0 |
| 4 and 8 versus 5–7 | Compare nonmercuric fungicides from company 1 to company 2 | 0 | 0 | 0 | 3 | –2 | –2 | –2 | 3 |
| 4 versus 8 | Compare fungicides from company 1 | 0 | 0 | 0 | 1 | 0 | 0 | 0 | –1 |
| 5 versus 6–7 | Compare older formulation (5) with newer formulations (6–7) from company 2 | 0 | 0 | 0 | 0 | 2 | –1 | –1 | 0 |
| 6 versus 7 | Compare the two new formulations from company 2 | 0 | 0 | 0 | 0 | 0 | 1 | –1 | 0 |

```
 test 5*(2.trt+3.trt) - 2*(4.trt+5.trt+6.trt+7.trt+8.trt) = 0

 (( 1)  5*2.trt + 5*3.trt - 2*4.trt - 2*5.trt - 2*6.trt - 2*7.trt
- 2*8.trt = 0

       F(  1,     35) =  152.88
            Prob > F =    0.0000

 test 2.trt - 3.trt = 0

 ( 1)  2.trt - 3.trt = 0

       F(  1,     35) =   17.67
            Prob > F =    0.0002

 test 3*(4.trt+8.trt) - 2*(5.trt+6.trt+7.trt) = 0

 ( 1)  3*4.trt - 2*5.trt - 2*6.trt - 2*7.trt + 3*8.trt = 0

       F(  1,     35) =   29.12
            Prob > F =    0.0000

 test 4.trt - 8.trt = 0

 ( 1)  4.trt - 8.trt = 0

       F(  1,     35) =    2.83
            Prob > F =    0.1016
```

```
 test 2*5.trt - (6.trt+7.trt) = 0

( 1)  2*5.trt - 6.trt - 7.trt = 0

      F(  1,     35) =    2.12
           Prob > F =    0.1542

 test 6.trt - 7.trt = 0

( 1)  6.trt - 7.trt = 0

      F(  1,     35) =    0.01
           Prob > F =    0.9051
```

### Built-in Multiple Range Tests

In many cases, and some statisticians think in most cases, the specific treatment comparisons should be planned in advance. Frequently, the experiment and treatments will indicate the planned comparisons you should look at. For example, an entomologist may be interested in how the current standard insecticide compares with new materials. These new insecticides, in addition, may have different modes of action and the researcher may wish to compare these different modes of action. This kind of information before the experiment is conducted determines what the planned comparisons will be.

There are, however, legitimate cases, I believe, where planned comparisons are not possible. Variety trials are a good example. The comparisons of interest in such a case can encompass all possible comparisons. More than likely, however, the comparisons of interest will depend on the individual viewing the information. I do variety trials that are distributed widely to growers, seed companies, and other researchers. Each has its own comparisons of interest. Growers may be interested in comparing their current variety to improved or better-performing varieties. Seed companies may be interested in comparing their varieties to their competitors and researchers could have a wide range of interests in the trial as it relates to their work. As the number of comparisons increases, the chance of committing a Type I error increases. For example, with 10 varieties, there are 45 possible pairwise comparisons. The comparisonwise Type I error can be calculated, for example, at the 5% level as $45 \times 0.05 = 2.25$, which is rounded to the nearest whole number, 2 in this case. This means there is the chance of finding two significantly different comparisons when, in fact, there are none. All possible comparisons together are often referred to as a family of comparisons and the Type I error rate in this case as the familywise error rate.

There are several methods available to evaluate all the pairwise comparisons. Load the dataset Onion Small Trial 1999.dta, which is a small onion variety trial. Then analyze the data with the **anova** command. Follow this with the **pwcompare** command. The commands and output are shown below:

```
anova yieldacre entry rep
pwcompare entry, pveffects
```

```
                  Number of obs =         15     R-squared     =  0.9178
                  Root MSE      = 78.4047        Adj R-squared =  0.8561

         Source |  Partial SS    df        MS             F      Prob > F
        --------+----------------------------------------------------
          Model |  548744.502     6   91457.4169       14.88      0.0006
                |
          entry |   532527.14     4   133131.785       21.66      0.0002
            rep |  16217.3618     2   8108.68091        1.32      0.3198
                |
       Residual |    49178.37     8   6147.29625
       ---------+----------------------------------------------------
          Total |  597922.872    14   42708.7765

Pairwise comparisons of marginal linear predictions

Margins      : asbalanced

------------------------------------------------------------
             |                                 Unadjusted
             |   Contrast    Std. Err.          t    P>|t|
-------------+----------------------------------------------
       entry |
      2 vs 1 |      78.65    64.01717        1.23    0.254
      3 vs 1 |  -10.40599    64.01717       -0.16    0.875
      4 vs 1 |   -372.438    64.01717       -5.82    0.000
      5 vs 1 |   -339.768    64.01717       -5.31    0.001
      3 vs 2 |  -89.05599    64.01717       -1.39    0.202
      4 vs 2 |   -451.088    64.01717       -7.05    0.000
      5 vs 2 |   -418.418    64.01717       -6.54    0.000
      4 vs 3 |   -362.032    64.01717       -5.66    0.000
      5 vs 3 |   -329.362    64.01717       -5.14    0.001
      5 vs 4 |      32.67    64.01717        0.51    0.624
------------------------------------------------------------
```

The first command (**anova**) generates the typical ANOVA table. The second command (**pwcompare** entry, **pveffects**) makes all pairwise comparisons between the varieties (entry), and the option **pveffects** indicates that the effects table should show the p values for the comparisons. The p values are listed in the last column where we can see that the first variety is different from varieties 4 and 5, but not different from varieties 2 and 3. The **pwcompare** command offers several other methods of computing a multiple range test. Each uses a different approach, which is discussed more fully later in the chapter. For example, enter the following command and see the output:

```
pwcompare entry, pveffects tukey
Pairwise comparisons of marginal linear predictions

Margins      : asbalanced

---------------------------
             |  Number of
             | Comparisons
-------------+-------------
       entry |          10
---------------------------

-----------------------------------------------------
             |                                Tukey
             |   Contrast   Std. Err.        t    P>|t|
-------------+---------------------------------------
       entry |
      2 vs 1 |      78.65   64.01717      1.23   0.737
      3 vs 1 |  -10.40599   64.01717     -0.16   1.000
      4 vs 1 |   -372.438   64.01717     -5.82   0.003
      5 vs 1 |   -339.768   64.01717     -5.31   0.005
      3 vs 2 |  -89.05599   64.01717     -1.39   0.649
      4 vs 2 |   -451.088   64.01717     -7.05   0.001
      5 vs 2 |   -418.418   64.01717     -6.54   0.001
      4 vs 3 |   -362.032   64.01717     -5.66   0.003
      5 vs 3 |   -329.362   64.01717     -5.14   0.006
      5 vs 4 |      32.67   64.01717      0.51   0.984
-----------------------------------------------------
```

The results between the unadjusted and Tukey's probabilities are similar. The same pairs of variety means are declared significant in each test ($P<0.05$). This will not always be the case, however, and can be seen by the different probabilities listed. For example, the first row (2 vs. 1) in the unadjusted comparison has a probability ($P>|t|$) of

0.254 and, in the Tukey comparison, it is 0.737. In both cases, there is no difference (P>0.05), but the probability is much higher with Tukey's test, which is known for being particularly conservative in declaring differences significant and this is reflected in the higher P value.

Several other multiple comparison tests are available in Stata. In addition to the unadjusted and Tukey's mentioned above using the **pwcompare** command, there are the Bonferroni's, Šidák's, Scheffe's, Student–Newman–Keuls' (SNK), Duncan's, and Dunnett's tests. Dunnett's is a special case of multiple range testing where one treatment is compared to all others. This might be used, for example, where a particular standard animal ration is compared to several other ration formulations. The treatment with the lowest identifying number is considered the standard for comparisons.

Post hoc multiple range tests are available in Stata as we have seen above. These tests are also included with CRD (completely randomized design) within the **oneway** command. Open the file virustestinoc.dta, which is the absorbance reading of virus-infected watermelon germplasm. Antigen/antibody testing can be highly specific for detecting virus diseases in plants. Such colorimetric tests rely on the absorbance of light at a specific wavelength to determine if an infection is positive. After loading the dataset, enter the command

```
oneway absorb trt, bonferroni tabulate
```

This results in the following output:

```
       1-5: |
  Different |
      virus |
 inoculated |
 watermelon |  Summary of ELISA Absorbance value
  germplasm |        Mean   Std. Dev.       Freq.
------------+------------------------------------
     PI 025 |       0.520       0.036          11
     PI 026 |       0.506       0.033          11
   PI 261-1 |       0.275       0.048          11
     PI 528 |       0.237       0.025          11
       Egun |       0.147       0.054          11
------------+------------------------------------
      Total |       0.337       0.156          55
```

```
                        Analysis of Variance
    Source              SS         df      MS            F     Prob > F
------------------------------------------------------------------------
Between groups    1.22886109    4    .307215272     185.11   0.0000
 Within groups   .082981547    50   .001659631
----------------------------------------------------------------------
    Total        1.31184263    54   .024293382

Bartlett's test for equal variances:   chi2 (4) =   6.7369
Prob>chi2 = 0.150

              Comparison of ELISA Absorbance value
    by 1-5: Different virus inoculated watermelon germplasm
                             (Bonferroni)
Row Mean-|
Col Mean |      PI 025       PI 026     PI 261-1      PI 528
---------+---------------------------------------------------
  PI 026 |      -0.014
         |       1.000
         |
PI 261-1 |      -0.245       -0.231
         |       0.000        0.000
         |
  PI 528 |      -0.283       -0.269       -0.038
         |       0.000        0.000        0.340
         |
    Egun |      -0.372       -0.359       -0.127       -0.090
         |       0.000        0.000        0.000        0.000
```

The **oneway** command offers three different multiple comparison tests as options. This includes Bonferroni, Scheffé, and Šidák. All three are presented in the Results window as a triangular matrix of the differences between the treatment means and a probability that these differences are significant. All three use different approaches to control the familywise Type I error rate. The **oneway** command is limited to a single factor model, which is not applicable except for a CRD. In agriculture, other designs, particularly RCBD (randomized complete block design) for field experiments, are much more common.

For example, the Bonferroni adjustment divides the selected probability by the number of comparisons and declares only those significant at this new probability level. So, for example, with the absorbance data above, there are 10 possible comparisons; therefore, the Bonferroni adjustment at the 5% level of significance would be

$$\frac{0.05}{10} = 0.005$$

You can see that, as the number of comparisons increases, the chosen probability quickly becomes very small. With 10 varieties and 45 possible comparisons, the 5% probability is now actually 0.001.

Šidák's adjustment uses the following formula to determine the probability at which the difference should be declared significant:

$$a = 1 - (1 - \alpha)^{1/n}$$

Again using the absorbance data with 10 possible comparisons and wishing to use a 5% level of significance, the new probability level would be

$$0.005 = 1 - (1 - 0.05)^{1/10}$$

Scheffé's approach is to calculate a multiplier ($S$), which then is multiplied against the standard error and this value then is used as the minimum difference for significance. This multiplier is calculated as

$$S = \sqrt{(t-1)F_{\alpha,\,(t-1),\ error\ df}}$$

where $t$ is the number of treatments and $F$ is the critical value often available in the $F$ distribution table in the back of statistics textbooks. The $S$ value can be easily calculated and displayed in Stata. Enter the following command to display the $S$ multiplier:

```
display sqrt(4*invFtail(4,50,0.05))
```

This also can be displayed by entering the following command immediately after calculating the ANOVA. The **oneway** command saves several scalars in r(), which can be viewed with **ereturn list**. Enter the following to calculate the $S$ value:

```
display sqrt(r(df_m)*invFtail(r(df_m),r(df_r),0.05))
```

S, which is 3.1982365 in this case, is then multiplied with the standard error of the difference between two means. The $SE$ value is calculated as

$$SE = \sqrt{s^2\left(\frac{1}{n_a} + \frac{1}{n_b}\right)}$$

where $s^2$ is the residual mean square or mean square error and $n_a$ and $n_b$ are the number of replications for treatments *a* and *b*. This can be calculated and displayed within Stata for our example with

```
display sqrt(.001659631*2/11)
```

Using the scalars from the ANOVA, it also can be displayed with

```
display sqrt(r(rss)/r(df_r)*2/(r(N)/(r(df_m)+1)))
```

Multiplying *S* by *SE* (0.01737098) results in 0.0556, which is then used as the minimum value to compare any two means. If the difference between the means exceeds this value, then the difference is considered significant at the specified probability level.

### Programming Scheffé's Test

Stata offers a wide variety of post hoc multiple range tests that can list the probabilities of all pairwise comparisons. However, results are rarely, if ever, presented in this format in the agricultural literature. It is more common to present these results with means followed by letters where any means having the same letters are not considered significantly different at the chosen probability level (usually 0.05 or 0.01).

To develop your programming skills further and develop programs that present these results in a more table-friendly format do-files of several of these multiple range tests have been developed. You should have already read the previous chapter on programming to help understand this process. We will be using the Strontium.dta dataset. This is a dataset of strontium levels found in various lakes (Zar, 1974, p. 152). Load the dataset and open the Do-file scheffe.do.

The scheffe.do program calculates the Scheffé's multiple contrasts test and presents the results as a list of means followed by letters indicating which means differ. Means followed by the same letter are not significantly different at the 5% level. This program assumes the experiment is an RCBD. The significance level and the experimental design, however, can be easily changed within the program. Remember, to use the program, first it must be run. This can be accomplished by selecting the Run or Do buttons on the top right of the Do-File Editor.

The program is heavily commented to explain how it works, but, in any case, I would like to walk through the program and explain exactly what it does. This basic format can be used with any of the multiple range tests with small changes to the calculations. There are three arguments entered with the **scheffe** command representing the dependent variable (depend), the independent or treatment variable (indep), and the replication variable (rep). The number of arguments can be changed for more complex designs. The next statement calculates the ANOVA and, again with more complex designs, this statement can be modified to accommodate such changes. The next statement sets the macro trt to the number of treatments from the saved scalars, e(df _ 1) plus 1. **Set more off** turns off the page pausing that can occur in the Results window when the output is particularly long.

The next two statements calculate the critical *S* value discussed above and the standard error as described above. These two are then multiplied together. This value is then the critical value in comparing any two means. The next couple of lines display the results of this calculation.

The **preserve** command is used to remember the dataset in memory. This is used in conjunction with the command **restore**, which occurs later in the program. The dataset is going to be collapsed to capture the treatment means in the next statement and then the dataset will be restored to its original form. You can see the effect of **preserve** and **restore** by having the Data Editor visible and entering the following three commands in the Command window:

```
preserve
collapse (mean) stron, by (lake)
restore
```

Collapsing the dataset by the treatments (indep), calculating the treatment means (depend), and sorting these results give us the treatment means in ascending order. These values then can be used later in the program.

Do-files can do more than just calculations and display results. Stata's programming language, like all computer programs, has the ability to both loop through statements many times as well as make decisions based on specific criteria. The next several statements in the program illustrate looping. The **forvalues** statement begins the loop. The format for this command is

```
forvalues x = #1(#d)#2 {
          statements...
}
```

The first number (*#1*) is the beginning value for x in the loop. The second value (*#d*) indicates how much x should be incremented or decremented, and the final value (*#2*) is the final value for the loop. An example is

```
forvalues x = 1(1)10 {
display `x'
}
```

This program loops from 1 to 10 in increments of 1 and displays the incremented value each time the loop is executed. This can be entered interactively in the Command window to see the results. When incrementing from the first value to the last value in units of 1, there is a shortcut method of entry.

```
forvalues x = 1/10 {
```

There are a couple of other criteria required by **forvalues** loops. The **{** brace must appear on the same line as the **forvalues** command and the **}** brace must appear on a line by itself. Frequently, while programming, particularly when using nested loops, it is easy to forget to enter the final brace. Stata's do-file editor can check and let you know which open and closed braces match. To see this, double click on any brace and the Do-File Editor will indicate its match.

Beginning with

```
forvalues z = `trt'(-1)2 {
```

there are two loops, one nested inside the other. The first loop begins with the number of treatments (5 with the Strontium.dta dataset) and loops down to 2. The next statement stores the calculated value of each mean minus the critical comparison value (*S*) in the macro `` `z' `` where z indicates each comparison, of which there are four with the Strontium.dta. The next **forvalues** statement loops from 1 to the number of treatments and is nested inside the previous loop. This means that for each loop of the **forvalues** z loop, the **forvalues** i loop is completed. For the Strontium.dta, this means the inside loop (**forvalues** i) is executed five times before the next time the **forvalues** z loop is executed. The inside loop (**forvalues** i) has a decision statement within it. Decision statements have the general format as covered in the previous chapter:

```
if exp {
     additional commands
}
else {
     additional commands
}
```

The **if** command evaluates the following expression to determine if it is true or false. If the statement is true, then the commands within the braces are executed; otherwise they are skipped. The **else** command is not necessary unless there are alternate commands to execute. As with the **forvalues** statement, a { brace must appear on the same line as the **if** statement and there must be a corresponding } brace on a line by itself.

This decision statement

```
if `test`z'' < `depend'[`i'] {
```

tests whether the value `` `test`z'' `` is less than the `` `depend'[`i'] `` value. If the value is lower than the macro `` v,`z' `` is incremented by one, which represents the number of means that do not differ. It may be worth it at this point to take a closer look at the values in this loop. Remember, macros that appear in open and closed quotes return the value stored there. So, for example, with `` `test`z'' ``, we are dealing with test5, test4, test3, and test2 as calculated values from the statement

```
local test`z' = `depend'[`z'] - `S'
```

Because `depend` is one of the variables from our dataset, to access the individual values in this variable, the [] brackets are used to identify the individual values. Remember, we have collapsed the dataset, thus, with the Strontium.dta, we are only dealing with five means.

The next series of statements begins with

```
local v1 = `trt'
```

which sets the first value of `v1` to the number of treatments, since there is no value for this macro because the previous loop stops with z equal to 2. The **`forvalues`** f loop again loops from 1 to the number of treatments. The first statement in this loop sets g to one less than the value of f, which in the first instance through the loop would set g to 0. The next if statement

```
if "`v`f''" != "`v`g''" {
```

compares the v value from the previous **`forvalues`** z loop one at a time to the value just preceding it. If they are not equal, t is incremented by one. The t value represents the number of letters to use when building the table of mean differences. With this **`if`** statement, we see the use of double quotes where single quotes are used for numerical values and double quotes for text values. Because the first value of vg is nothing, the comparison between vf and vg must be as text to prevent an error in the program.

The next series of statements begins with

```
local j1 = 1
```

which sets the first value of `j` to `1`. This is followed by a couple of other macros (`h` and `p`) declared and their values set. This is followed by another loop (**`forvalues`** `f`), which begins with the number of treatments and counts down to 2 in increments of 1. Again the value of v is compared with an **`if`** statement that compares the value of v just preceding in the sequence. If these values are not equal, then the starting point is calculated and stored in jh and the ending point in kp. Once the beginning and ending points are calculated for all the means that are similar, then letters can be accumulated in the macro alphf. This is done in the **`forvalues`** b loop, which loops the number of letters required as stored in t from above. Finally, the letters are displayed next to the appropriate mean (depend) in descending order. The program then restores the dataset back to its original state.

This program represents one method of solving this problem and in programming there are often several different methods that can be used to arrive at the same solution. In programming, it is often the marginal case that gives the most trouble. For example, in the statement

```
local alph`f' = "`alph`f''" + substr("abcdefghijklmnop
qrstuvwxyz",`b',1)
```

the number of letters available includes the entire alphabet. What if, although unlikely, the number of differences exceeds this? This condition could result in an error stopping the program. Programs that are to be distributed as ado files and act like a built-in function often require a great deal of programming to handle these marginal cases. Programs that only you will use won't require this kind of rigor. It is important and cannot be emphasized enough, however, the need to document your code as you build it and use macro names that give some idea of what it is. The Scheffé Test is considered a rather conservative test that many consider overprotecting against type II errors (accepting the null hypothesis when the alternate hypothesis is true).

Duncan's New Multiple Range Test (MRT) was developed by the statistician David Duncan in 1955 (I guess the *New* could be dropped). Actually the New was added to distinguish this test from a previous one proposed by Duncan. Duncan's MRT is a modification of the Student-Newman-Kuel test that adjusts the alpha level based on the distance of the treatment means from each other. Unlike the Scheffé Test, which is often considered too conservative, Duncan's MRT is often considered too liberal in declaring two means as different. Unlike Scheffé Test that uses a single value to compare the treatment means, Duncan's MRT uses different values to compare treatment means based on how far apart the treatment means are when ranked in descending order. So, for example, 10 treatments where the highest treatment mean was compared to the fourth largest treatment mean would use a different value for comparison than the highest value compared to the sixth largest mean.

Load the dataset watertrial2007frtchar2.dta and open the do-file, duncan.do. We will be using these files to examine Duncan's MRT.

This do-file requires an ado file that is not part of the official package of Stata ado files. The ado file in question is qsturng. To find this program, enter in the Command window while connected to the Internet

```
findit qsturng
```

This command will open a Viewer window with a list of several Stata Technical Bulletins (STB). Download dm64, which is in STB-46, and install this ado command. **qsturng** stands for q studentized range, which is found in tables at the back of many statistics textbooks. This command requires three inputs: the number of treatments, error degrees of freedom, and probability rate. For a 5% probability, enter the value 0.95 and for a 1% probability enter 0.99.

The duncan do-file is almost identical to the scheffe do-file in terms of presenting the results. It differs in the comparison values used to compare means. With the Scheffé Test, a single test value is used, whereas with Duncan's MRT there is a different value for each comparison. Look at the segment of code below to see how this is calculated.

```
local var = (e(rmse))^2              // Error mean square from ANOVA
local repl = e(df_2)+1               // Number of replications

local sd = sqrt(2*`var'/`repl')      // Standard error of the mean
                                        difference

local trt = e(df_1)+1                // Number of treatments

forvalues x = 2/`trt' {              // Loop from 2 to number of
                                        treatments

quietly: qsturng `x' e(df_r) 0.95^(`x'-1)

/* Calculates the Studentized Range (this function is not part of
the official Stata ado files and will have to be downloaded */

local stu`x' = $S_1                  //qsturng saves its results in
                                       global S_1 & $S_1 returns the
                                       value

local r`x' = `stu`x''*`sd'/sqrt(2) /* Using the Studentized Range
                                       to calculate the significant
                                       difference based on rank order
                                       distance */
        }                            // End 'x' brace
```

From the previous ANOVA, several e() scalars are available to use in the program. The e(rmse) is the root mean square error, which when squared results in the error mean square or variance. This value along with the number of replications can be used to calculate the standard error of the mean difference. The first loop uses the **forvalues** loop, which loops from 2 to the number of treatments and is used to calculate the Studentized Range value with the **qsturng** command. This then is used to calculate the comparison values based on the rank difference.

With the duncan.do and watertrial2007frtchar2.dta files loaded, make sure the duncan.do file has been executed, which can be done from the Do-File Editor window by clicking the Run or Do icons in the upper right corner on a Macintosh or the same icons in the icon bar on a Windows computer (see Chapter 7, Figure 7.1). Once this is done, enter the following command:

```
duncan length trt rep
```

This will result in the following output:

```
         Number of obs =      51     R-squared     =  0.9276
         Root MSE      = .743141     Adj R-squared =  0.8708

     Source |  Partial SS    df       MS           F     Prob > F
  ----------+----------------------------------------------------
      Model |  198.193996    22   9.008818       16.31     0.0000
            |
        trt |  195.604525    19  10.294975       18.64     0.0000
        rep |  2.25708993     3  .752363311       1.36     0.2746
            |
   Residual |  15.4632584    28  .552259229
  ----------+----------------------------------------------------
      Total |  213.657254    50  4.27314509

Duncan's Multiple Range Test (P≤0.05)
17.37     a
12.56     b
12.12     bc
12.05     bc
11.88     bcd
11.62     bcde
11.38     bcde
11.28     cde
11.19     cdef
11.19     cdef
11.15     cdef
11.10     cdef
11.00     cdefg
10.75     defg
10.73     defg
```

```
10.51      efg
10.00      fgh
9.84       gh
8.88       hi
8.71       i
```

The **duncan.do** command outputs the ANOVA table with the means in descending order followed by letters signifying significant differences. Means followed by the same letter are not significantly different at the 5% level. To see how Duncan's MRT differs from Scheffé's Test load and run the **scheffe.do** command with the same data.

In addition, there are do files for calculating both Tukey's *w* procedure (**tukey.do**) and Student–Newman–Kuels' Test (**snk.do**) available online.* Each uses a slightly different approach to determining significant differences.

Finally, there is a test for comparing a single treatment to all other treatments called Dunnett's test. This has obvious applications in agriculture, such as comparing a standard variety to new introductions or a common pesticide to new formulations. Several years ago, I was interested in using this test. It had been proposed that this test could be used as a Multiple Comparison with the Best (MCB). This involved ranking the means in descending or ascending order (depending if you were comparing to the largest value or the smallest value) and choosing the top-ranked mean as the standard to compare all other means. I was interested in comparing onion pungency data in this fashion. I was not able to find Dunnett's *q'* values implemented in Stata, so I called the company and they informed me that this was not available. Several months later, I received an ado-file from Stata that calculated Dunnett's *q'*. Dunnett's can now be accessed within the **pwcompare** command, and the **dunnett** command is available from the Web. Just type **findit dunnett**. Open the dataset virustestinoc.dta, which as mentioned before is a dataset of watermelon germplasm screened for resistance to zucchini yellow mosaic virus, Egyptian strain (ZYMV-E). Values recorded for each entry are light absorbance values from the colorimetric virus test. Enter the following command and see the results:

```
dunnett absorb trt, control(5)
```

* Files available online at http://www.crcpress.com/products/isbn/9781466585850.

```
-------------------------------------------------------------------------
         |                                                    Different
   trt   |  Mean      Diff   [ 2-Sided  95% SCI ] abs(Diff)  from Control?
---------+---------------------------------------------------------------
Egun     |.147471      ---       ---         ---        ---          ---
PI 025   |.5197166 .3722456 .3281233    .4163679  .3722456          Yes
PI 026   |.506014  .358543  .3144207    .4026653  .358543           Yes
PI 261-1|.2749171 .1274461 .0833238    .1715684  .1274461          Yes
PI 528   |.2370599 .0895889 .0454666    .1337112  .0895889          Yes
-------------------------------------------------------------------------
     Diff = mean (trt)-mean(control)
     Different from mean (control) if abs(Diff) >  .044122
```

Turning off the value labels will show the numbers coding for the different entries or enter **label list** to see the numbers and value labels. Egun is coded as 5, so that is why it is listed as the **control** in the command line. As you can see from the output, Egun has a significantly lower absorbance value compared to the other entries.

# 9 Preparing Graphs

## Graphing in Stata

One of Stata's strengths is its capability of generating publication-quality graphs. These graphs can be easily exported in a number of formats that can be incorporated into other files or saved as stand-alone output. It is beyond the scope of this text to cover all of the graphing capabilities of Stata, so I will concentrate on a few graph types and the editing features available.

There is a Graphics menu available from which graphs can be constructed. Several of the graphs available in Stata are listed as separate items under this menu. These graphs are some of the most common types as well as several that are available for diagnostic purposes. Diagnostics of commands like **regress** are often more easily seen and understood when viewed graphically. In addition, several graphs are available in more than one location. For example, under the Graphics menu, the Distributional graphs submenu lists several diagnostic graphs, which are also available under the Distributional plots and tests, which is under the Summaries tables and tests submenu under the Statistics menu.

All the graphing commands can be entered in the command window beginning with the command **graph**. For example, open the dataset Large Onion Dataset 2001-02.dta. This is a dataset of an onion variety trial conducted in the winter of 2001–2002 with 31 varieties. Enter the command and see the results:

```
graph bar (mean) yield, over(variety,
label(angle(vertical)))
```

It may be easier, particularly as you begin using graphs, to select from the menus and fill in the various dialog boxes. This may be

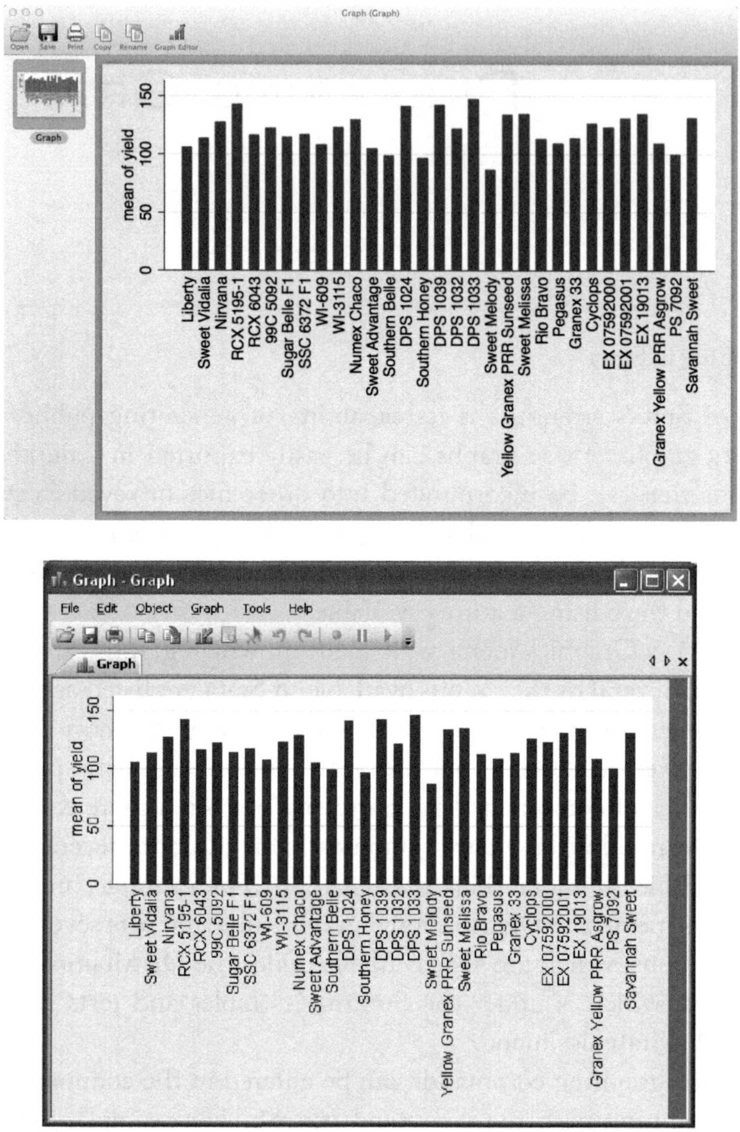


**Figure 9.1** Graph window in Stata with onion variety trial yields represented in a bar graph on a Macintosh (above) and Windows computer (below).

particularly useful when trying to construct a graph for the first time. As you get comfortable with the graphing features and if you have specific routine graphs to construct, the Command window may be more useful and quicker. The menu item for this graph is the Bar chart item under the Graphics menu.

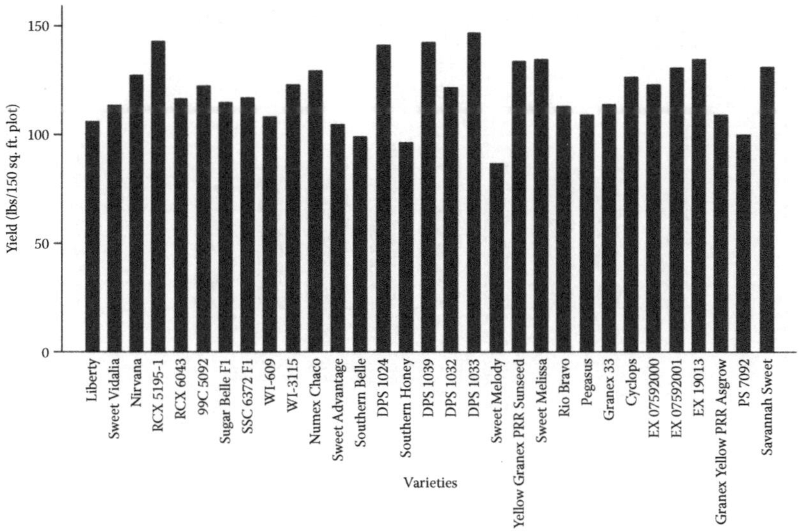


**Figure 9.2** Onion variety trial bar graph formatted for readability and output as a tiff file.

To make the above graph more readable, the font sizes were made smaller, the label for the *y*-axis was changed to be more descriptive, and the *x*-axis label Varieties was added. The actual graph as it first appeared in the Graph window is shown in Figure 9.1. Notice the differences as compared to Figure 9.2.

The layout of the Graph window between Macintosh and Windows computers is somewhat different, but the overall functionality is the same. This is particularly evident when the graph is in the edit mode.

At the top of the Graph window (Figure 9.1) are several icons for opening, saving, printing, copying, and renaming graphs. These icons act as expected allowing the user to quickly handle these functions. The next icon at the top of this graph is the Graph Editor icon. Selecting this icon places the current graph in an editing mode where all of the various options and styles can be incorporated into the graph. Such changes were made to the graph to produce the output in Figure 9.2.

Figure 9.3 shows the Graph window after the Graph Editor icon has been selected. In this mode, various elements of the graph can be selected and changed. For example, double clicking on the list of varieties opens a dialog box where various aspects of the *x*-axis can be changed. The detail of control is very good, but may be unfamiliar for

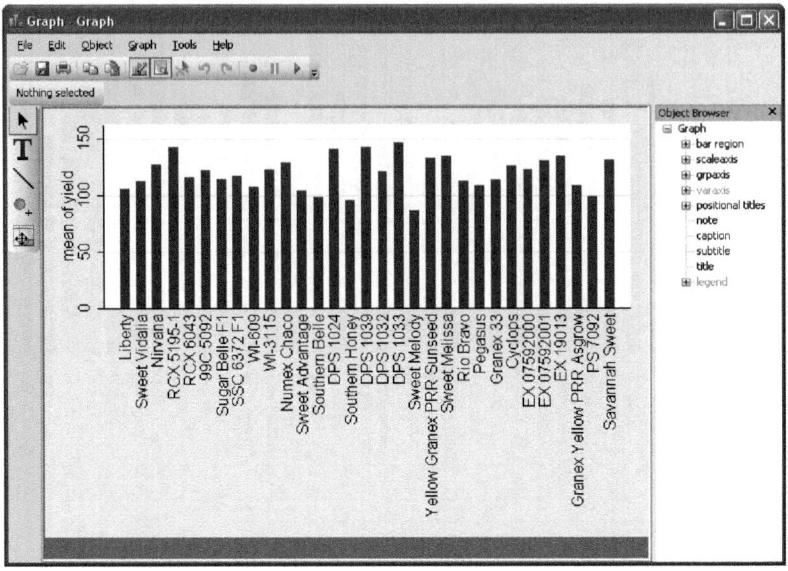


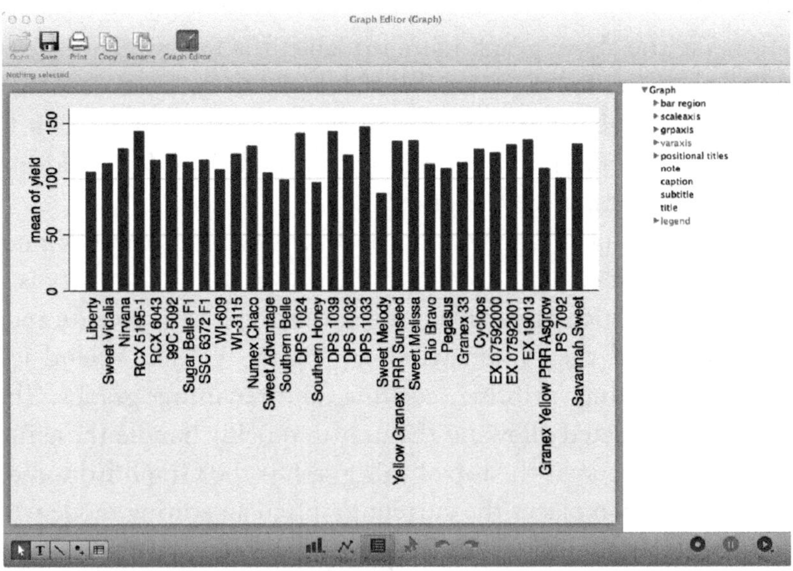


**Figure 9.3** Graph window in editing mode for a Windows computer (above) and Macintosh (below).

those used to other graphing-capable programs, where font sizes and spacing are changed based on point or line spacing. This difference is minor, however, after you have gotten used to it. With the window in Graph Editor mode, the Data Editor window is no longer available as are several items under the Graphics menu.

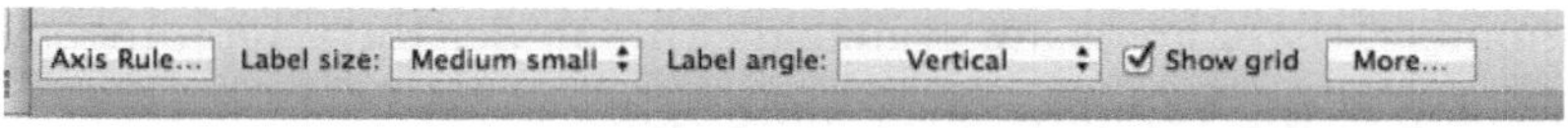


**Figure 9.4** Items available when the scaleaxis is selected on a Macintosh.

**Figure 9.5** Bottom of the Graph Editor window showing several icons for editing and changing graphs.

On the right side of the Graph Editor window is the graphing elements list. Selecting these items will place a marque (red rectangle) around the specific graph element. In addition, the nothing-selected region of the editor will change to show specific details of the element. Figure 9.4 shows what is available if the scaleaxis element is selected from the right side of the window (this places a red marque around the *x*-axis). Double clicking on a graphing element in the list will bring up a dialog box, which can be used to make changes to that element. Items, such as the *x*-axis scale, label size, angle, and grid, can be easily accessed and changed. In addition, selecting the More... button opens a dialog box with a complete set of options for this axis.

At the bottom of the Graph Editor window are several additional icons. These include, on the lower left side of the window, icons for selecting items, text entry, adding lines, adding marks and related objects, and grid editing (Figure 9.5). These items will appear on the upper left side of the window on Windows computers. Once you have added text, lines, or marks to a graph, reselect the selection arrow to select these items for additional editing. To edit an added object, double click the item with the selection arrow for a dialog box of available editing options.

At the bottom center of the Graph Editor window are several additional icons on a Macintosh computer. The first icon labeled Graph can be used to quickly access several aspects of the graph and make changes. This icon is used to change such things as titles, graph size, and aspect ratio, to name a few. The next icon, labeled Object, can be used to lock and unlock various elements of the graph as well as show and hide selected graph items. The next icon turns the sidebar on and off. The next icon allows deselection of a selected item. Finally, the last two icons are for undoing and redoing the previous action. There does not appear to be any limit to the number of undos. On Windows computers, these items are available either under the Graph menu or at the top of the Window (see Figure 9.3).

Finally, on the lower right part of the Graph Editor window on Macintosh computers are three buttons (Record, Pause, and Play) that are used to record a sequence of changes to a graph that can be saved and used later. These icons are available at the top center of the window on Windows computers (see Figure 9.3). This can be particularly useful if you have several graphs to create that will be similar in appearance and detail.

Many of the listed graph types under the Graphics menu are particularly useful for evaluating a dataset to meet certain underlying criteria like normality. Some of these graphs also can be helpful in exploring relationships between variables. And, finally, many of the listed graphs are some of the most frequently used. The Bar chart, Dot chart, and Pie chart are commonly used and are self-explanatory. I have illustrated the use of the Bar chart above.

The Histogram item under the Graphics menu constructs a histogram or frequency bar graph with the data's frequency within a category represented by the height of the bar. The number of bars can be controlled either as continuous where the number of bars or bins are specified or discrete where each individual value is represented. Open the dataset Large Onion Dataset 2001-02.dta and enter the following:

```
histogram yield, normal
```

This constructs a histogram of onion yields and superimposes the normal density function over the histogram. Once this graph has been constructed it can be edited by selecting the Graph Editor icon at the top of the window (Figure 9.6).

Select the Graph Editor icon to put the graph in editing mode and select the Record button in the lower right of the window on a Macintosh or the same button in the upper center on a Windows computer. Change the color of the bars, add a title, subtitle, change the background color to white, and change the *x*-axis label (Figure 9.7). Once these changes have been made, click the Record button again to stop recording and save the file. Just such a file was created called Histogram Defined.grec and can be used over and over again.

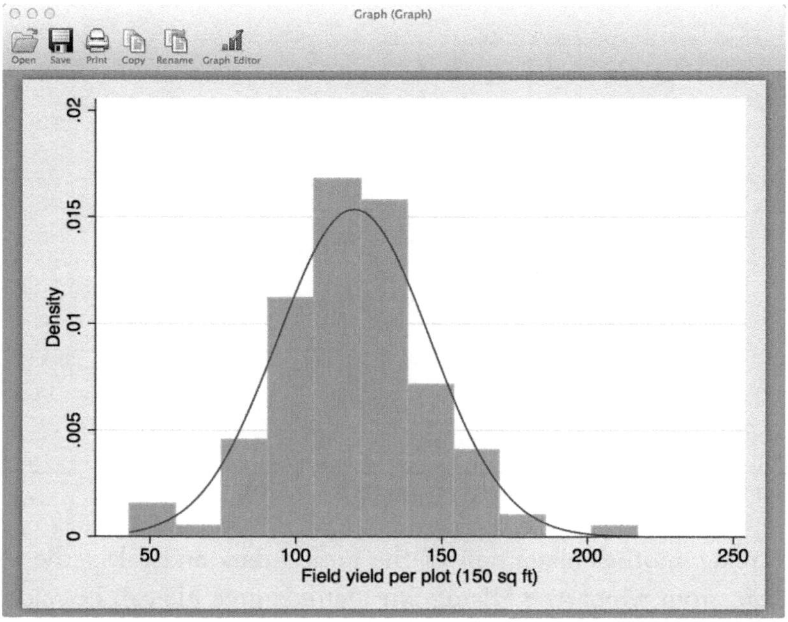


**Figure 9.6** Histogram of yield data from Large Onion Dataset 2001-02 with a normal density function superimposed over it.

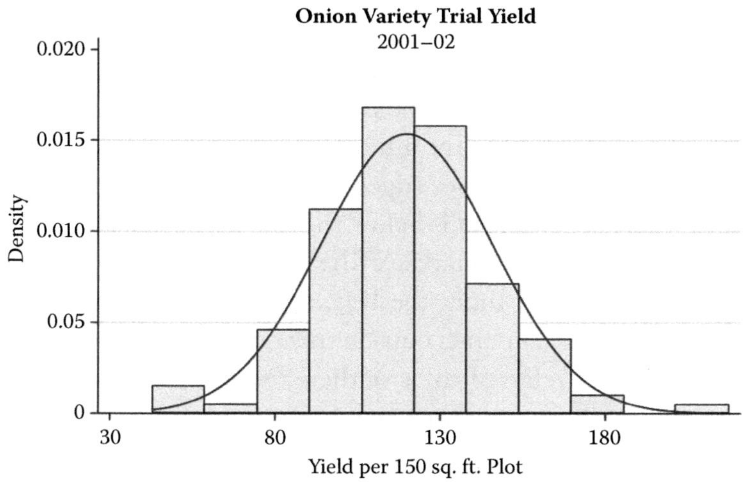


**Figure 9.7** Histogram of yield data with several editing changes.

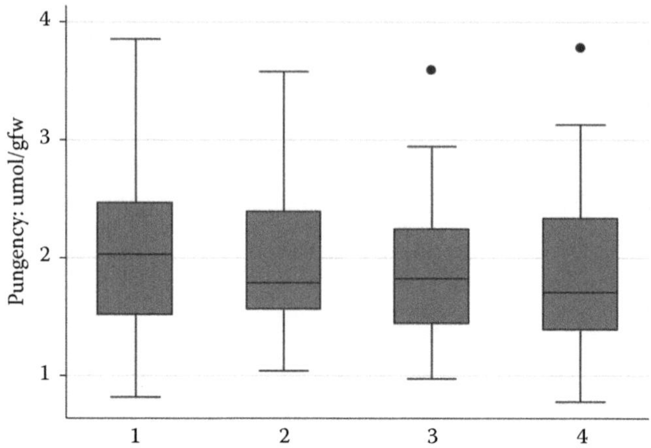


**Figure 9.8** Box plots of onion pungency grouped by replications.

Re-create another histogram of the jumbo data and select the Play button, from which the Histogram Defined.grec file can be selected and played back. This creates a histogram of jumbo (≥3 in.) onion yields with the editing changes from the Histogram Defined.grec file.

The next item under the Graphics menu is the Box plot, which constructs box plots or what are sometimes called *box* and *whisker* plots. These simple diagrams offer a wealth of information about the sample. Select Box plot under the Graphics and construct box plots for pungency with rep as the grouping variable under Categories. These are box plots of onion pungency grouped by replications (randomized complete block design, RCBD) (Figure 9.8). This illustrates the kind of information presented in this type of graph.

The box represents 50% of the data and is often referred to as the *interquartile range* (IQR). The line in the middle of the box is the median. The lower and upper edges of the box are the 25% and 75% quartiles where 25% of data is below this value (25% quartile) or 25% is above this value (75% quartile). The whiskers represent the upper and lower range or 1.5 times the IQR above and below the median, whichever is less. Data points outside this range are marked individually and are often referred to as outliers. Medians that are near the bottom edge of the box indicate the data are skewed to the right and medians near the top of the box are skewed to the left.

The Scatterplot matrix item under the Graphics menu allows you to look at the relationship between different variables (Figure 9.9).

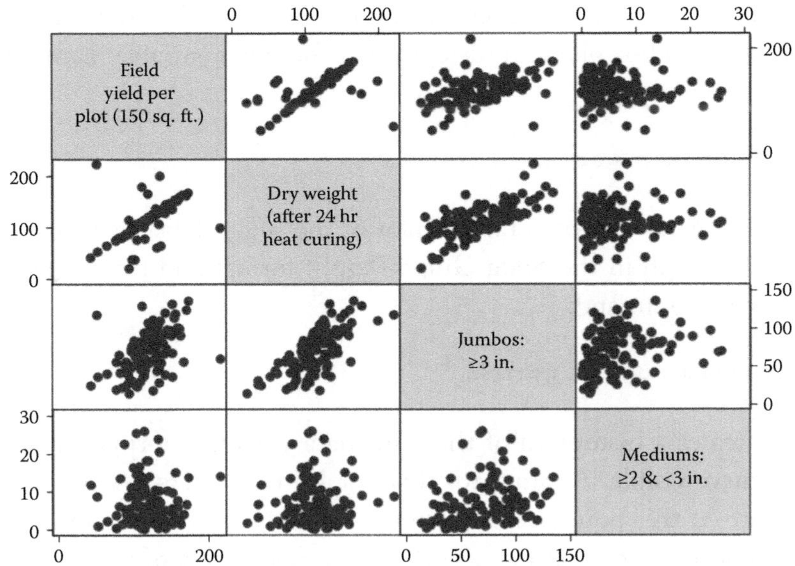


**Figure 9.9** Scatter plot matrix of onion yield components.

Again using the Large Onion Dataset 2001-02, enter the following command and see the results:

```
graph matrix yield drywts jumbo mediums
```

In this scatter plot matrix, it is easy to see some fairly strong relationships between field yield (weights immediately after harvest) and dry weights (weights after 24 hours of heat curing). This would be expected because the dry weights are just slightly less than the field weights. There also appears to be a relationship between yield and jumbo (≥3 in.) onions, but not much of a relationship between yield and mediums (≥2 and <3 in.).

There are several other graph types to choose from under the Graphics menu. Many of these are used for specific statistical analyses. For example, Regression diagnostics plots are used to evaluate regression analysis and are covered in Chapter 10 (Correlation and Regression).

The table of graphs under the Graphics menu is used to combine one or more graphs into a single file. This can be helpful when several graphs are related in some fashion and together they enhance the presentation. This feature also can help relating two different graphs presenting the same data. Using the Large Onion Dataset 2001-02, create both a box plot and histogram of the sugar variable. Enter first the command

```
graph hbox sugar
```

This creates a horizontal box plot of the sugar data. At this point, save the graph in the Stata Graph (*.gph) format and then enter the following command:

```
histogram sugar, normal
```

This creates a histogram of the same data with the normal distribution curve visible. Again, save this graph in the Stata Graph (*.gph) format. At this point, because computer path names are going to be different on each machine, use the menu item Table of graphs under the Graphics menu. Select the Browse... button to select your graphs. First, select the histogram and click Accept. Next, do the same for the box plot graph. Next, select the Options button at the top of the dialog box and under Layout: select Columns from the drop-down menu. Below this are the number of columns, which should be 1. Finally, click the OK button. This will create a graph with both the histogram and box plot together, one above the other. You may notice that the *x*-axis for both graphs does not line up. This can be corrected by selecting the Graph Editor button and then double clicking the *x*-axis of the box plot. In the dialog box that opens, select the Scale button and click the box Extend range of axis scale. At this point, you will be adjusting the Lower limit (< = 6.8). You can try various values to see how the scale on the box plot *x*-axis changes. A value of 6.4 appears to line up the two axes for the histogram and box plot (Figure 9.10).

The last two items on the Graphics menu are Manage graphs and Change scheme/size. The Manage graphs has several subitems, which allow for the management of graphs in memory including changing their names, copying, dropping, describing, and changing the graph in memory. Changing the scheme or size lets you quickly change the overall look and size of the graphs in memory. More details about graphing will continue in the next chapter on correlation and regression.

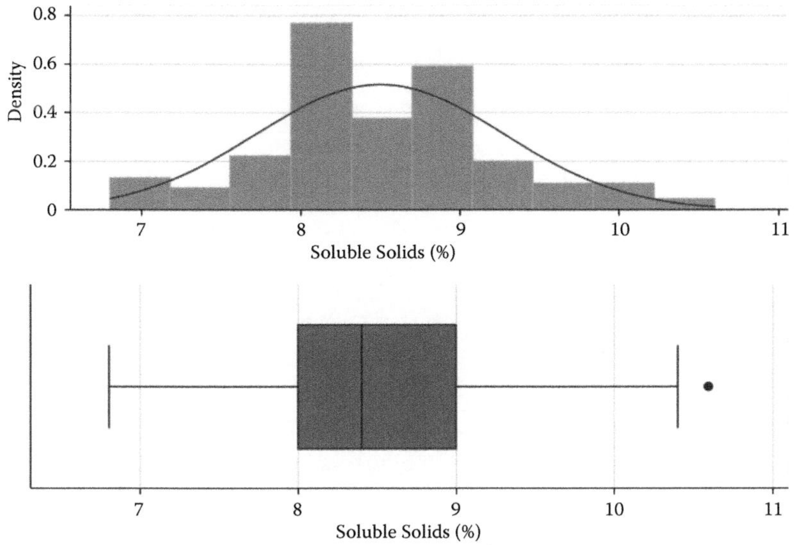


**Figure 9.10** Combined graphs of onion soluble solids (sugar) data with a histogram and box plot.

# 10
# Correlation and Regression

### Correlation

Statistical correlation is a method of determining if two sets of data can be related to each other. For example, the thickness of tree rings can be correlated with the amount of rainfall. So, in wet years, tree rings appear larger than in dry years. In this particular case, there appears to be a causal relationship—more rainfall (water) results in more growth. Correlations do not have to be positive; there can be a negative correlation between two variables. For example, a commodity price may go down as the supply increases. In many such cases, the link between the variables is obvious. A correlation between an increase in fertilizer and an increase in yield is another example where this link is obvious. Oftentimes, however, correlations occur where there is no apparent causal relationship. A rather famous example is foot size in children and spelling ability. There is a strong positive correlation between these two items. The reason there is such a good correlation between the two is not because foot size increases spelling ability. The association has to do with the child's age and level of education. This is an important point about correlation; it is not a cause and effect relationship. Correlation in this context can help identify relationships that might ultimately be a cause and effect relationship. They also can be unrelated as in the foot size and spelling ability. Therefore, interpretation of correlation results should be approached with caution, particularly if there is no obvious mechanism for the two to be linked.

The most common linear correlation is often called simple correlation, total correlation, or product-moment correlation. This type of correlation is measured by the coefficient of correlation and is designated by $r$, which is an unbiased estimate of $\rho$ (Greek rho). Correlation is a unit-independent measure of association. There is a related technique called *regression*, which is often confused with correlation. As a general rule of thumb, regression is used where the independent variable (X) is fixed, such as in a planned experiment, while with correlation there is no such constraint. Situations where the independent variable (X) is random is often referred to as a model II, while fixed effect models are considered model I; for example, using different rates of fertilizer to see what effect it has on yield in a replicated study. Correlation, on the other hand, can be evaluated between any two sets of data.

Correlation is defined as

$$r = \frac{Covariance\ (X, Y)}{\sqrt{Variance\ X\ Variance\ Y}}$$

$$r = \frac{\sum (X - \bar{X})(Y - \bar{Y}) / (n-1)}{\sqrt{\sum (X - \bar{X})^2 / (n-1)}\sqrt{\sum (Y - \bar{Y})^2 / (n-1)}}$$

$$r = \frac{\sum (X - \bar{X})(Y - \bar{Y})}{\sqrt{(X - \bar{X})^2 (Y - \bar{Y})^2}}$$

*r* will be a value between –1 and 1 with values close to –1 or 1 indicating a high degree of correlation whether negative or positive. Load the dataset Hog Price Data.dta, which is a dataset of marketed hogs and price per hundred weight (Little and Hills, 1978, p. 172). Let's begin by graphing the price of hogs against hogs marketed. To do this, select the Twoway graph (scatter, line, etc.) under the Graphics menu. This will bring up a dialog box (Figure 10.1) where a new graph can be constructed.

In this dialog box, select the Create… button to create a new graph. This will open another dialog when the default plot (Scatter) is selected. Do not change anything in the dialog box, but select the price variable from the Y variable: drop down and hogs as the X variable: drop down. Then select the Accept button. Finally, select the OK button in the Twoway dialog. This will create a scatter plot graph of these variables as shown in Figure 10.2. This can also be created with

```
twoway (scatter price hogs)
```

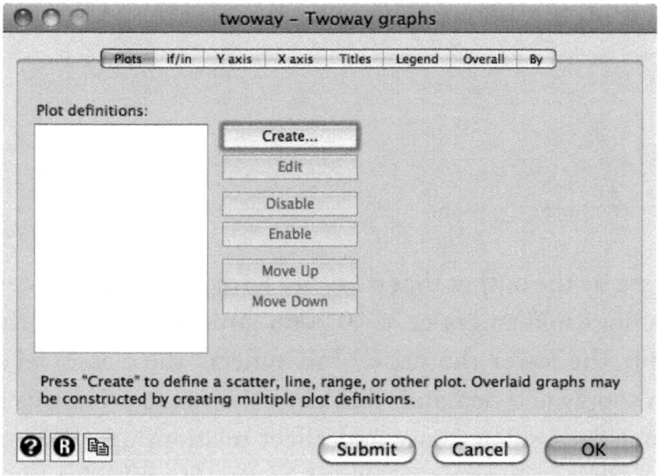


**Figure 10.1** Twoway graphing dialog box on a Macintosh computer.

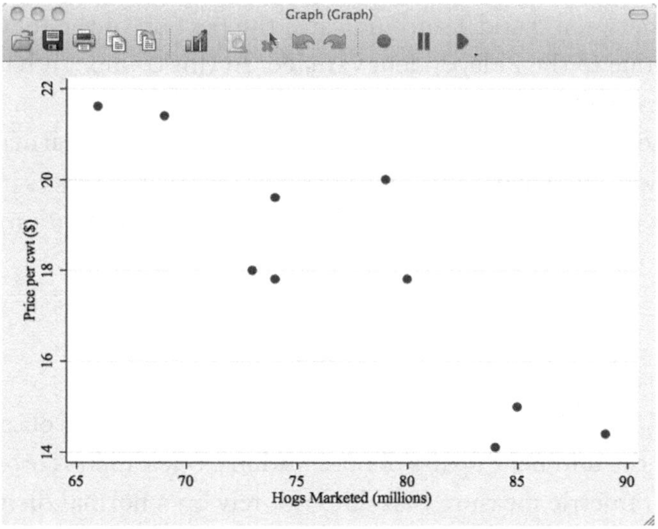


**Figure 10.2** Scatter plot graph of hog prices and number marketed.

The dots in the scatter plot appear to be trending downward as you move from left to right on the *x*-axis. This suggests a negative correlation with lower prices associated with more hogs marketed.

Now let's see what the correlation between these two variables is. Enter the command

```
correlate price hogs
```

This results in the following output:

```
(obs=10)

       |    price      hogs
-------+------------------
 price |   1.0000
  hogs |  -0.7068    1.0000
```

We can see by the output that there is a fairly high negative correlation between hogs sold and price at –0.7068. This suggests that the higher the supply the lower the price. This reflects the classic relationship between supply and demand. Remember, however, that this relationship is not an absolute cause and effect relationship. There could be conditions where prices are high and hogs sold are high as well. There are usually more factors affecting a market than price or supply alone.

Squaring $r$ results in the coefficient of determination ($r^2$). This value will be between 0 and 1 and indicates the portion of the total sum of squares due to the independent variable. In this context, it has important consequences in regression analysis.

Another method of calculating a correlation is Spearman's rank correlation coefficient. This method relies on the differences in rank of the data points. The formula for Spearman's rank correlation is

$$r_s = 1 - \frac{6\sum d_i^2}{n(n^2 - 1)}$$

where $d_i$ is the difference between ranks for each pair of observations and $n$ is the number of pairs of observations. Spearman's correlation is a nonparametric measure that does not rely on a normal distribution. Spearman's correlation also occurs on a scale of –1 to 1 with values close to these showing a high degree of correlation. Load the dataset Nicotiana correlation.dta. This is a dataset of flower measurements from a Nicotiana cross (Steel and Torrie, 1980, p. 276). First enter the following command:

```
correlate tube limb base
```

This results in a matrix of correlations between the three variables. Now enter the following command:

```
spearman tube limb base
```

This also results in a matrix of correlations, but the values are slightly different. In addition, with the **spearman** command, when only two variables are included, the probability is shown indicating whether the correlation is significant. The results of the correlations and the **spearman** command with tube and limb are shown below:

```
correlate tube limb base

(obs=18)
             |     tube      limb      base
-------------+------------------------------
        tube |   1.0000
        limb |   0.9550    1.0000
        base |   0.7972    0.6781    1.0000


spearman tube limb base
(obs=18)

             |     tube      limb      base
-------------+------------------------------
        tube |   1.0000
        limb |   0.9611    1.0000
        base |   0.7525    0.6767    1.0000


spearman tube limb

 Number of obs =        18
Spearman's rho =          0.9611

Test of Ho: tube and limb are independent
    Prob > |t| =          0.0000
```

Whether calculating simple correlation or Spearman's correlation, the relationships are considered linear. Not all relationships are linear and real associations can occur and not be linear.

### Linear Regression

Linear regression is often referred to as a model I problem because the independent variable is fixed. This means that, in general, the independent variable (X) has been decided on in advance or has some finite value. For example, you might be interested in regressing food consumption of hens based on body weight. Body weight in this context is fixed because there is a finite range for hen weights.

As you might remember from algebra class, the formula for a line is Y = a + bX, where a represents the y-intercept and b is the slope. It is this equation that we are trying to fit with the data. Trying to fit a line to data points that best represents where those data points occur is the purpose of linear regression. The criterion for determining this line is to minimize the sum of squares of these deviations. This method is often referred to as the least-squares method. The formula for estimating the slope (*b*) is

$$b = \frac{\sum (X - \bar{X})(Y - \bar{Y})}{\sum (X - \bar{X})^2}$$

$\hat{Y}$ is an estimate of Y defined as

$$\hat{Y} = \bar{Y} + b(X - \bar{X})$$

Open the dataset Hen Regression.dta, which is a small dataset of leghorn hen weight and food consumed over 350 days (Steel and Torrie, 1980). Enter the command

```
regress food weight
```

This will result in the following output:

```
Source    |       SS          df       MS           Number of obs =      10
----------+------------------------------           F(  1,      8) =   16.23
 Model    |  90.8354996        1   90.8354996       Prob > F       = 0.0038
Residual  |  44.7685335        8   5.59606669       R-squared      = 0.6699
----------+------------------------------           Adj R-squared  = 0.6286
   Total  |  135.604033        9   15.0671148       Root MSE       = 2.3656

-------------------------------------------------------------------------------
    food |    Coef.    Std. Err.    t     P>|t|    [95% Conf. Interval]
---------+---------------------------------------------------------------------
  weight |  7.690104   1.908736   4.03   0.004    3.288551     12.09166
   _cons |55.26328     9.534895   5.80   0.000    33.27578     77.25079
-------------------------------------------------------------------------------
```

The top portion of the output is similar to an ANOVA (analysis of variance) table. This is because regression and analysis of variance are very similar. The input with the **regress** command assumes that the independent variable, `weight`, is continuous. This results in a single degree of freedom for the Source Model. The remainder of this table is similar to an ANOVA table. There is no F-test or probability reported in the ANOVA table because this information is presented elsewhere in this output. The F-test and probability are presented along with the $R^2$, adjusted $R^2$, and the mean square error (MSE), which is the square root of the Residual mean square (MS). The $R^2$ is the square of the correlation coefficient discussed previously and represents that portion of the total sum of squares (SS) that is the Model SS. This value (90.8354996/135.604033) is 0.6699. The closer the $R^2$ value is to 1 the better the model fits the data. The adjusted $R^2$, as mentioned in Chapter 5, is an adjustment to the $R^2$ and in this context does not have much meaning.

The bottom portion of the output lists several pieces of information, the most important of which is the coefficients (Coef.). The value for weight (7.690104) is the slope of the least squares estimate of the linear equation for these data. The _cons (55.26328) represents the Y-intercept. Substituting a hen weight between 4.4 and 5.9 lbs in the equation and including the slope, it is possible to predict food consumption. It is important to understand that substituting a hen weight is only valid within the range of actual weights. The regression line is invalid beyond this range because the function may be quite different outside these numbers. This makes sense, particularly in this context, if you think about it. For example, plugging in a 100-lb hen to find out its food consumption makes no sense because there is no such thing as a 100-lb chicken (at least I'm pretty sure there isn't, Foghorn Leghorn not withstanding).

After conducting the regression analysis, it may be worthwhile to examine the results to determine if the underlying assumptions are correct. One of these assumptions is the residuals occur randomly and independently of the underlying model. Stata has two commands that show whether this is true, the **rvpplot** and **rvfplot**. The former plots the residuals against the predictor or X value in the linear regression. The latter plots the residuals against the fitted or Y value.

Residuals are the differences between the actual data and the prediction expected from the model. To see these graphs, a regression must first be performed, then these commands can be entered. With the Hen Regression.dta in memory, enter the following:

```
regress food weight
```

Then enter

```
rvfplot, yline(0)
```

and then

```
rvpplot weight, yline(0)
```

The first command is to ensure that a regression has been calculated; otherwise the next two lines will result in an error message. The next command graphs the fitted versus residual data and, finally, the **rvpplot** graphs the predictor (weight) versus the residuals. The **yline (0)** places the red horizontal line on the graph to make the results more readable. Figure 10.3 shows these results. Stata can only display one graph window at a time; therefore, as each new graph command is entered, the graph window shows those results. Once a graph is saved it can be opened at the same time the Graph window is displaying the other graph. Graphs appear, then, on the left side of the Graph window.

These graphs should have their plotted values occurring randomly around 0 on the *y*-axis. If there were a pattern to these data points, then it would indicate that the residuals were not random and independent. Although these plots appear very similar, with more complex regressions, with more than one independent variable, these graphs will appear different from each other.

Let's look at another dataset that shows an example where these residuals show such a pattern. Load the dataset Rice Varieties Regression.dta. This is a dataset of tiller numbers and yield for two different rice varieties (Gomez and Gomez, 1984, p. 373). Again run the regression (**regress** `yield tiller`) and plot the residuals against the predictor and fitted values (**rvfplot** and **rvpplot** `tiller`).

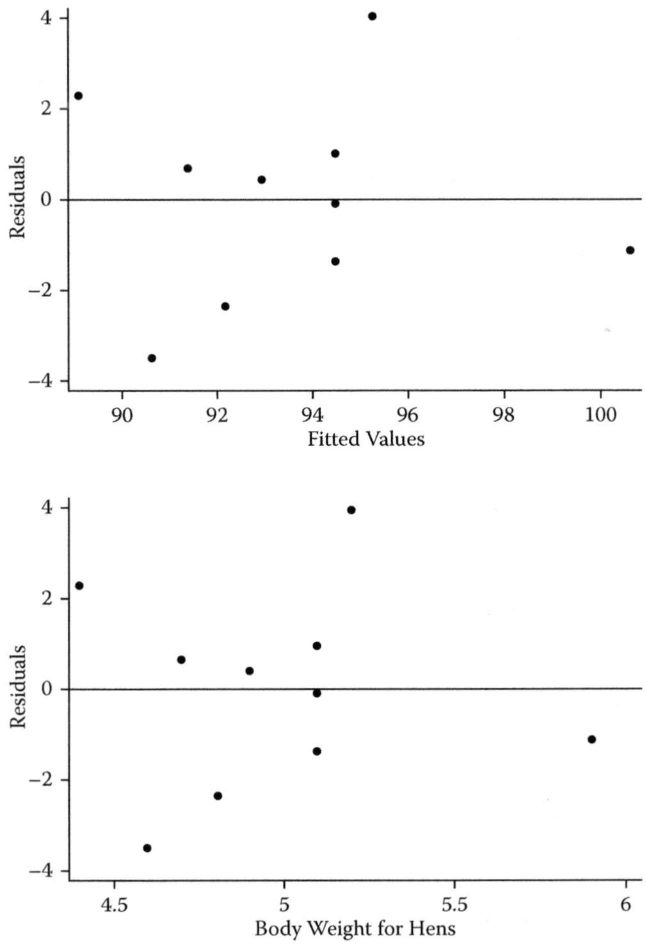


**Figure 10.3** The fitted value versus residuals and predictor versus residuals graphs for the hen data regression.

Figure 10.4 shows the fitted values versus the residuals (the predictor versus residuals would look the same with this model). Notice the points are above the 0 line near 4,500 and 7,000 on the *x*-axis and below the 0 line in the center. This dataset actually has two different varieties that perform quite differently. To see this graphically enter the following command:

```
twoway (scatter yield tiller in 1/8) (lfit yield
tiller in 1/8) (scatter yield tiller in 9/16)
(lfit yield tiller in 9/16)
```

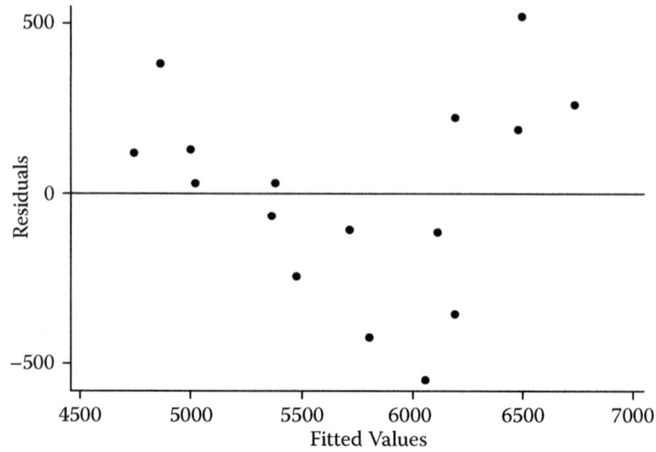


**Figure 10.4** Fitted values versus residuals for the rice tiller and yield data.

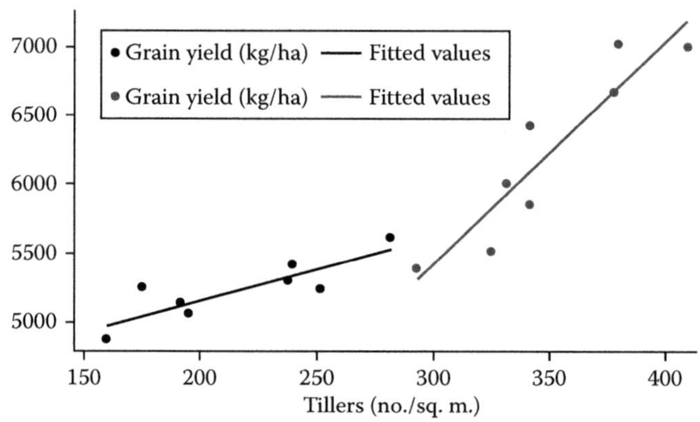


**Figure 10.5** Scatter plots and fitted lines for two rice varieties: Milfor 6(2) and Taichung Native 1.

Looking at Figure 10.5, it is quite obvious that there are two distinct varieties. Running the regression separately for each variety and examining the residuals will show they are independent and occur randomly for each. To run both regressions simultaneously, enter the following command:

```
by variety, sort : regress yield tiller
```

```
-------------------------------------------------------------------
-> variety = Milfor 6(2)

  Source |       SS         df        MS        Number of obs =        8
---------+----------------------------          F(  1,      6) =    16.04
   Model | 260252.056    1   260252.056         Prob > F       =   0.0071
Residual | 97377.9439    6   16229.6573         R-squared      =   0.7277
---------+----------------------------          Adj R-squared =   0.6823
   Total | 357630        7        51090         Root MSE       =    127.4

-------------------------------------------------------------------
   yield |  Coef.    Std. Err.    t    P>|t|    [95% Conf. Interval]
---------+---------------------------------------------------------
  tiller |4.555356  1.137575  4.00   0.007    1.771811     7.338901
   _cons |4242.127  250.6494 16.92   0.000    3628.809     4855.444
-------------------------------------------------------------------

-------------------------------------------------------------------
-> variety = Taichung Native 1

  Source |       SS        df         MS        Number of obs =        8
---------+------------------------------        F(  1,      6) =    36.16
   Model | 2463313.09    1    2463313.09        Prob > F       =   0.0010
Residual | 408730.407    6    68121.7344        R-squared      =   0.8577
---------+------------------------------        Adj R-squared =   0.8340
   Total | 2872043.5     7    410291.929        Root MSE       =      261

-------------------------------------------------------------------
   yield |    Coef.    Std. Err.    t    P>|t|   [95% Conf. Interval]
---------+---------------------------------------------------------
  tiller |16.01067    2.662517   6.01   0.001   9.495721     22.52561
   _cons | 620.014    937.1012   0.66   0.533   -1672.99     2913.018
-------------------------------------------------------------------
```

Although, in this case, the differences between these two varieties are quite obvious, it may not always be so. One area that may be of interest when examining such data is determining if the slopes or regression coefficients differ. In the Rice Varieties Regression.dta, the variety `variable` uses a value label to indicate the variety names; however, the actual coding is 0 for variety Milfor 6(2) and 1 for variety Taichung Native 1. This coding is important as we shall see in a moment. If the coding had been any other numbers, it would have to be replaced with 0 and 1. Now we will create a new variable called `taichung`, which is the product of variety and tiller. To do this, enter the following command:

```
generate taichung = tiller * variety
```

This results in a new variable, `taichung`, where all the entries for variety Milfor 6(2) are 0. Then a regression is run with the following command:

```
regress yield tiller variety taichung
```

This results in the following output:

```
  Source |       SS       df       MS              Number of obs =      16
---------+------------------------------           F(  3,     12) =   53.03
   Model | 6709577.4   3  2236525.8                Prob > F      =  0.0000
Residual | 506108.351 12 42175.6959                R-squared     =  0.9299
---------+------------------------------           Adj R-squared =  0.9123
   Total | 7215685.75 15 481045.717                Root MSE      =  205.37

------------------------------------------------------------------------------
   yield |    Coef.   Std. Err.     t     P>|t|    [95% Conf. Interval]
---------+--------------------------------------------------------------------
  tiller | 4.555356 1.833819     2.48   0.029    .5598086     8.550904
 variety |-3622.112 840.8033    -4.31   0.001 -5454.065    -1790.159
taichung | 11.45531 2.784214     4.11   0.001    5.389028     17.52159
   _cons | 4242.127 404.0575    10.50   0.000    3361.761     5122.492
------------------------------------------------------------------------------
```

Notice in the output that the coefficients for `tiller` and `_cons` are the same as the regression for variety Milfor 6(2). The coefficient for `variety` is the difference between the *y*-intercepts for the two varieties (620.014 – 4242.127 = –3622.112). Finally, the `taichung` coefficient is the difference in the slopes for variety Taichung Native 1 minus the slope for variety Milfor 6(2) (16.01067 – 4.555356 = 11.45531). The t value for `taichung`, which is 4.11, is a test to see if the slopes of the two varieties are significantly different and, in this case, they are (P>|t| = 0.001).

Frequently, there will be more data points of Y, the dependent variable (e.g., in a replicated study), than of the independent variable X. Usually when this occurs the dependent data points are averaged before the regression is calculated. This will eliminate noise or variability in the analysis. This additional data, however, can be useful. For example, in a variety trial, data will be collected from each replication. A variety trial is not generally analyzed with regression because the varieties are individual items, but the added data points may be helpful in examining other relationships. Load the dataset Onion Pungency Regression.dta. This is a dataset of onion pungency (i.e., the measurement of pyruvate as µmoles/gram fresh weight, which is an indirect indicator of how hot or pungent an onion is) for a replicated variety trial. This dataset could be used to examine the regression of days to harvest and its effect on onion pungency. Enter the following command:

```
regress pungency days
```

which results in the following output:

```
    Source |       SS       df       MS              Number of obs =     118
----------+------------------------------            F(  1,   116) =   19.40
     Model |5.40363379   1  5.40363379               Prob > F      =  0.0000
  Residual |32.3065373 116  .278504632               R-squared     =  0.1433
----------+------------------------------            Adj R-squared =  0.1359
     Total |37.7101711 117  .322309155               Root MSE      =  .52774

----------------------------------------------------------------------
  pungency |    Coef.    Std. Err.      t   P>|t|  [95% Conf. Interval]
----------+------------------------------------------------------------
      days |  -.0181154 .0041126 -4.40   0.000 -.026261    -.0099698
     _cons |   5.862495  .644764  9.09   0.000 4.585459     7.139531
----------------------------------------------------------------------
```

The results suggest that there is a significant effect of lower onion pungency with later harvest date. The $R^2$ value is relatively low suggesting that the relationship is rather weak. A further analysis can be done evaluating these data with days as both a continuous and categorical predictor. Enter the following command:

```
anova pungency c.days days
```

This results in the following output:

```
              Number of obs =       118       R-squared     =  0.4065
              Root MSE      = .447025         Adj R-squared =  0.3800

     Source |  Partial SS    df       MS           F     Prob > F
-----------+-----------------------------------------------------
      Model |  15.3290569     5  3.06581138     15.34     0.0000
            |
       days |  .521338564     1  .521338564      2.61     0.1091
       days |  9.92542313     4  2.48135578     12.42     0.0000
            |
   Residual |  22.3811142   112  .199831376
-----------+-----------------------------------------------------
      Total |  37.7101711   117  .322309155
```

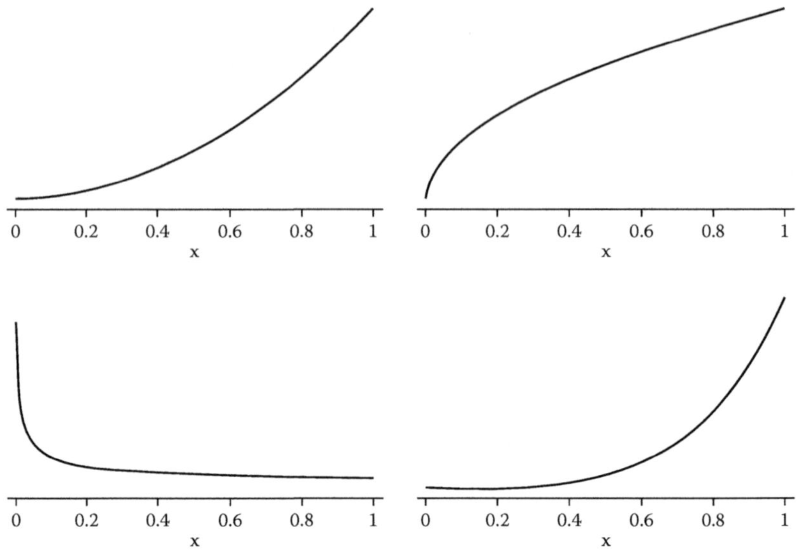


**Figure 10.6** Different types of curves. The top two are examples of power curves. The lower left is a decay curve and the lower right is a growth curve.

The ANOVA command allows there to be both continuous and categorical predictors in the same analysis. The `c.` before a variable tells the program to treat this variable as continuous, as it would be treated in a simple regression analysis. In addition, the `days` variable is entered as a categorical variable. This analysis suggests that the regression has continuous variable `days` not significant (P>F = 0.1091) while the categorical `days` variable is significant indicating the regression is not linear. The $R^2$ in this case is greater than the previous analysis suggesting this is a better fit.

Clearly ambiguous results such as this suggest more work should be done. Perhaps a single onion variety could be harvested over an extended period of time and analyzed for pungency to get a clearer understanding.

Not all relationships are linear, but may follow some other functional relationship. Figure 10.6 shows several graphs of functions that may be encountered in agricultural studies. Datasets that appear to follow one of these graphs may be analyzed with linear regression after the data have been transformed using a log transformation.

Open the dataset Onion Bulb Sizes.dta. This is a dataset from Little and Hills (1978, p. 199). Graphing this datum indicates an increase in weight with increasing bulb diameter [**twoway (scatter** weight diameter**)**]. Input the following command and see the results:

```
regress weight diameter

      Source |       SS          df        MS      Number of obs =    30
-------------+---------------------------------- F(  1,     28) =   503.26
       Model |  152562.215      1  152562.215 Prob > F        =   0.0000
    Residual |   8488.0869     28  303.145961 R-squared       =   0.9473
-------------+---------------------------------- Adj R-squared =   0.9454
       Total |  161050.302     29  5553.45868 Root MSE        =   17.411

------------------------------------------------------------------------
      weight |     Coef.   Std. Err.    t   P>|t|   [95% Conf. Interval]
-------------+----------------------------------------------------------
    diameter |4.143827   .1847158   22.43 0.000   3.765454    4.522201
       _cons |-138.2188 11.63165 -11.88 0.000 -162.0451 -114.3924
------------------------------------------------------------------------
```

The analysis suggests that the data have a significant linear fit with an $R^2$ of 0.9473. There are, however, some problems with this analysis. For one thing, the *y*-intercept is –138.2188, which means that as the bulb diameter gets below about 30 mm the bulb weights are negative. Obviously, this can't be so. In addition, the data points appear to be above the expected linear function with very low and very high bulb diameters (Figure 10.7). The data points should occur randomly above and below the predicted linear function. This type of data can often be explained with a power curve, which has the general equation of

$$Y = aX^b$$

Now, generate new variables with the following command:

```
generate lgdiameter = log10(diameter)
generate lgweight = log10(weight)
```

The new variables are the base 10 logarithmic transformation of the bulb diameter and weight data. The original data and the transformation are graphed below with a linear prediction line for each. Notice how the transformed data better fit a straight line.

Now, enter the following command with the transformed variables and see the results:

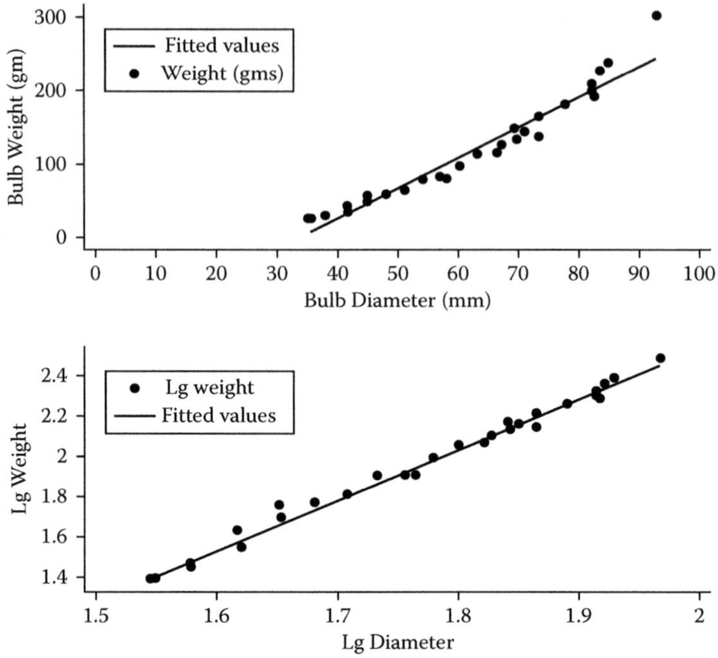


**Figure 10.7** Onion bulb diameter and weight data with the original data plotted in the top graph and the transformed data plotted in the bottom graph.

```
regress lgweight lgdiameter
```

```
   Source |       SS       df       MS              Number of obs =      30
----------+------------------------------           F(  1,    28) = 3128.25
    Model | 3.22008254     1  3.22008254            Prob > F      =  0.0000
 Residual |  .028821921   28  .001029354            R-squared     =  0.9911
----------+------------------------------           Adj R-squared =  0.9908
    Total | 3.24890446    29  .112031188            Root MSE      =  .03208
```

Although the results are similar with a significant linear function, the second table has an $R^2$ value of 0.9911, which is higher than in the first table and the y-intercept is closer to 0 at –2.486792. Because the analysis was done on transformed data, the results are the linear equation y = 2.511754 x – 2.486792. This should be transformed back to the original units by taking the antilog of this equation, which is $y = x^{2.511754} + 0.00325993$. The antilog of the constant –2.486792 is found by raising 10 to the power of this value. In Stata, if you had used natural logarithms for the transformation, the inverse of this would be **exp()**. Finally, enter the following command to graph the original data points with the new equation:

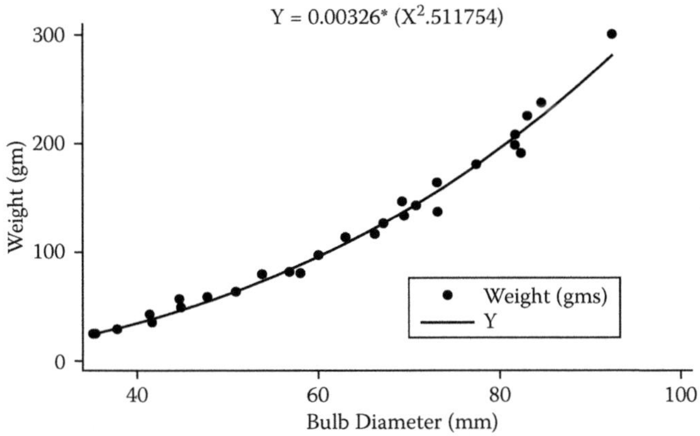


**Figure 10.8** Data points of onion bulb diameter and weight with regression line.

```
twoway (scatter weight diameter) (function y
=.00325993*x^2.511754, range (diameter))
```

The equation was added to the graph (Figure 10.8) to show what the Y function was. The graph also shows the entire number after X should be an exponent as follows:

$$Y = 0.00326 * (X^{2.511754})$$

Another type of curve that is often seen in agriculture is the exponential curve, which can be a growth or decay curve (see Figure 10.6). This type of curve will have the general form of

$$Y = aX^b$$

Open the dataset Cabbage Height.dta, which is a small dataset of plant height above the cotyledons measured on a weekly basis (Steel and Torrie, 1980, p. 456) (Figure 10.9). Graph this data with

```
twoway (scatter height week)
```

Now add a variable with a logarithmic transformation of the plant height (Figure 10.10).

```
generate logheight = log (helght)
```

In this case, only the plant height is transformed, not the weeks since this variable is already linear. Now, do the regression with the transformed plant height against weeks and plot the results.

```
regress logheight week
twoway (scatter logheight week) (lfit logheight week)
```

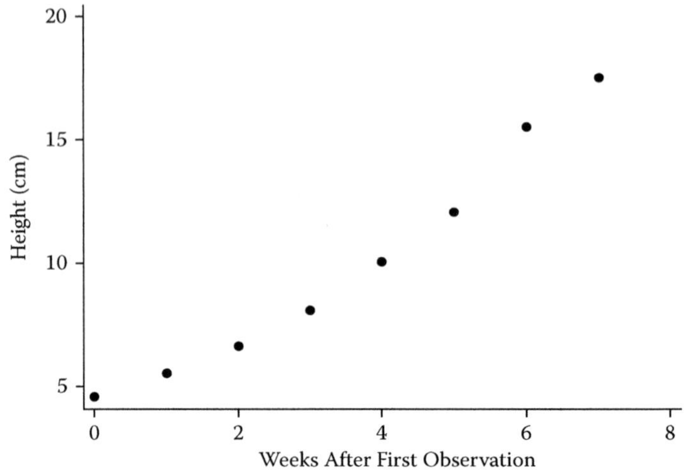


**Figure 10.9** Cabbage plant height measured on a weekly basis.

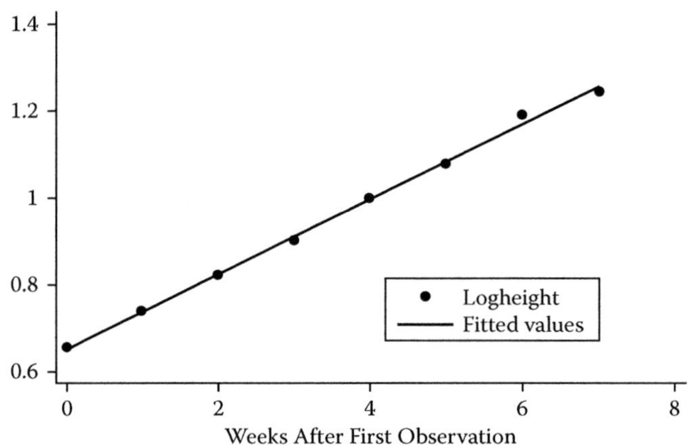


**Figure 10.10** Linear fit of logarithms of cabbage height data.

```
      Source |       SS       df       MS              Number of obs =       8
-------------+------------------------------           F(  1,      6) = 2650.51
       Model |  .313255845     1  .313255845           Prob > F      =  0.0000
    Residual |  .000709122     6  .000118187           R-squared     =  0.9977
-------------+------------------------------           Adj R-squared =  0.9974
       Total | .313964967      7  .044852138           Root MSE      =  .01087

------------------------------------------------------------------------------
  logheight | Coef.     Std. Err.    t    P>|t|     [95% Conf. Interval]
-------------+----------------------------------------------------------------
       week |.0863624  .0016775  51.48  0.000     .0822578    .0904671
      _cons |.6513264  .0070174  92.82  0.000     .6341554    .6684975
------------------------------------------------------------------------------
```

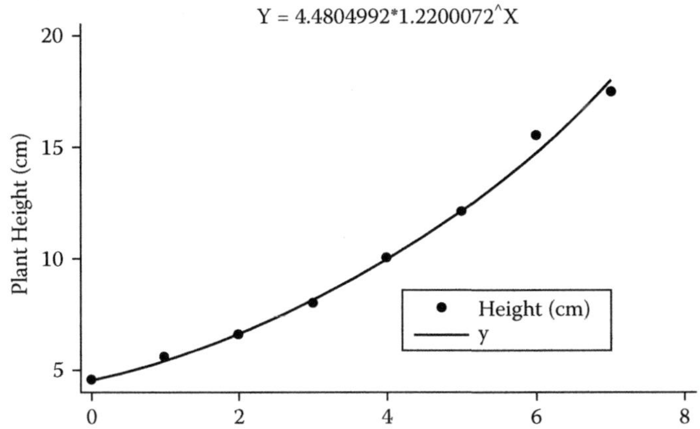


**Figure 10.11** Cabbage plant height fitted with exponential function.

The fit is very good with an $R^2$ value of 0.9977 (Figure 10.11). Taking the antilog of the linear equation results in Y = 4.4804992*1.2200072$^x$. The two numbers are 10 raised to the power of the coefficients (0.6513264 and 0.0863624). This can be plotted with the following entry:

```
twoway (scatter height week) (function y =
4.4804992*1.2200072^x, range (week))
```

Along with linear functions, there also can be functions that are referred to as polynomial functions that have the general expression of (Figure 10.11):

$$Y = a + bX + cX^2 + dX^3 + \ldots$$

These functions can have as many terms as one less than the total number of treatments. Usually the more terms the better the fit (greater $R^2$), but this can be misleading and difficult to interpret in a biological sense. The first term, $bX$, is referred to as the first-degree term and is nothing more than the linear function ($Y = a + bX$). The second term ($cX^2$) is the second-degree term or the quadratic equation ($Y = a + bX + cX^2$). The next is the third-degree term or the cubic equation and the fourth-degree term is referred to as the quartic equation. Usually the first, second, or third term equations are evaluated because there can be some biological basis for these. Higher order equations (i.e., 4, 5, 6, etc.) although possible to calculate are difficult or impossible to interpret in a biological or agricultural context.

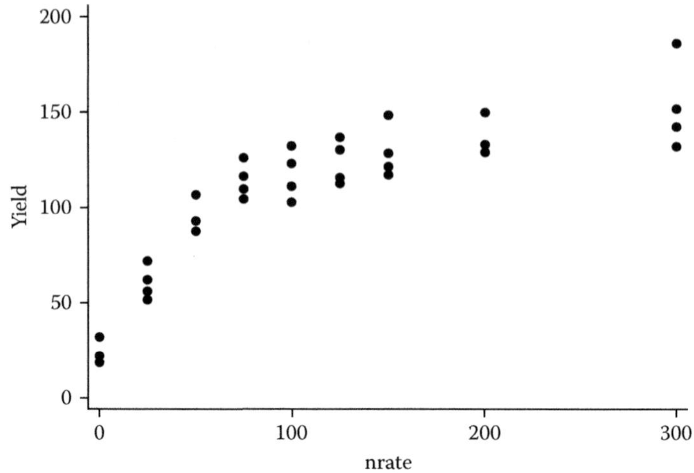


**Figure 10.12** Plotted data of onion yield based on nitrogen fertilizer rate.

Open the dataset Onion Fertility 2005.dta and look at the data for yield graphically with the following command:

```
Twoway (scatter yield nrate)
```

Look at the graph and the data appear to have a somewhat curved shape. This gives us a clue as to how the data should be handled (Figure 10.12). Enter the following regression command and look at the output:

```
regress yield nrate c.nrate#c.nrate

Source    |      SS        df        MS         Number of obs =       36
----------+--------------------------           F(  2,     33) =    99.40
   Model  | 48601.884       2   24300.942       Prob > F       =   0.0000
Residual  |8067.45567      33   244.468354      R-squared      =   0.8576
----------+---------------------------          Adj R-squared  =   0.8490
   Total  |56669.3397      35   1619.12399      Root MSE       =   15.635

------------------------------------------------------------------------
   yield  | Coef.      Std. Err.    t     P>|t|    [95% Conf. Interval]
----------+-------------------------------------------------------------
   nrate  |.9462218   .0945669   10.01   0.000   .7538241       1.13862
          |
c.nrate#
c.nrate   |-.0019585 .0003058   -6.40   0.000 -.0025806     -.0013364
          |
   _cons  | 38.53506 5.759268    6.69   0.000  26.81774      50.25238
------------------------------------------------------------------------
```

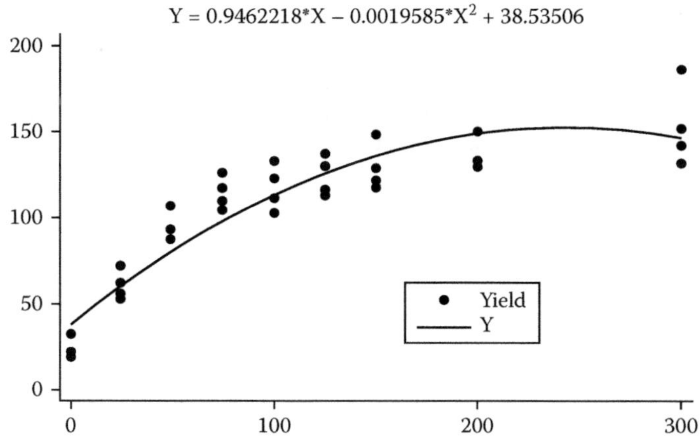


**Figure 10.13** Graph of onion yield with the corresponding quadratic equation.

The $R^2$ value is very good at 0.8576 and all three coefficients have a significant t-value. To see these results graphically, enter the following command and see the resulting graph:

```
twoway (scatter yield nrate) (function y =.9462218*x
-.0019585*x^2 + 38.53506, range(nrate))
```

Using the fit plots category for a quadratic equation with a confidence interval (**qfitci**) will result in the same curve with a confidence interval (Figure 10.13). The command to enter is

```
twoway (qfitci yield nrate) (scatter yield nrate)
```

For presentation purposes, you may wish to collapse the dataset averaging the yield data by nitrogen rate and then construct the graph. This makes for a cleaner presentation and does not detract from the results. The order in the command also is important; entering the **(qfitci** `yield nrate`**)** first ensures the quadratic curve and confidence interval appears behind the data points (Figure 10.14). Reversing the order, putting **(scatter** `yield nrate`**)** first hides some of the data points behind the confidence interval.

Additional power terms can be added, such as the third, fourth power, etc., and although the equation fit may improve it really doesn't add any more to the understanding of the underlying data. A yield curve such as this with increasing amounts of nitrogen fertilizer

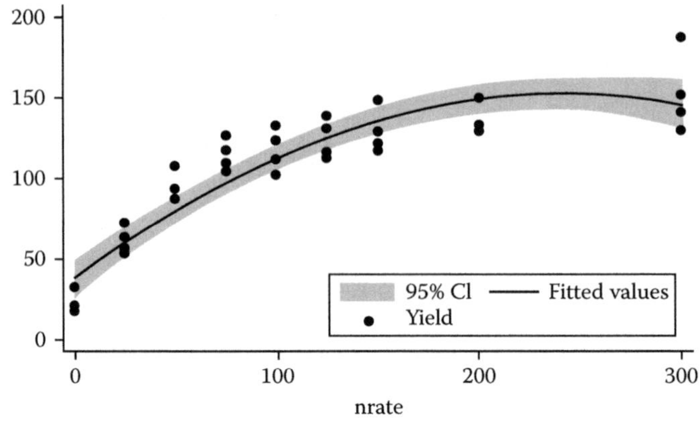


**Figure 10.14** Graph of onion yield with a quadratic curve and confidence intervals.

makes sense. You would expect an increase up to a point and then yields would drop off because of overfertilization.

Stata also offers the **nl** command that allows the analysis of nonlinear regression with any type of function. There are several built-in functions with this command that can be analyzed because they are so widely used. They include exponential regression with one asymptote, logistic function (symmetric sigmoid shape), and Gompertz function (asymmetric sigmoid shape).

Open the dataset Barley Yield.dta, which is a dataset measuring total dry matter based on days after seed drilling (Clewer and Scarisbrick, 2001, p. 99). Enter the following command and see the results:

```
nl log3: yield days
```

```
(obs = 19)

Iteration 0:  residual SS =  21.47574
Iteration 1:  residual SS =  7.036482
Iteration 2:  residual SS =  6.362144
Iteration 3:  residual SS =  6.354819
Iteration 4:  residual SS =  6.354655
Iteration 5:  residual SS =  6.354652
Iteration 6:  residual SS =  6.354652
Iteration 7:  residual SS =  6.354652

   Source |      SS       df       MS
----------+------------------------------Number of obs =       19
    Model |1757.32536     3  585.775119   R-squared     =   0.9964
 Residual |6.35465168    16  .39716573    Adj R-squared = 0.9957
----------+------------------------------ Root MSE      = .6302109
    Total |1763.68001    19  92.8252637   Res. dev.     = 33.10988
```

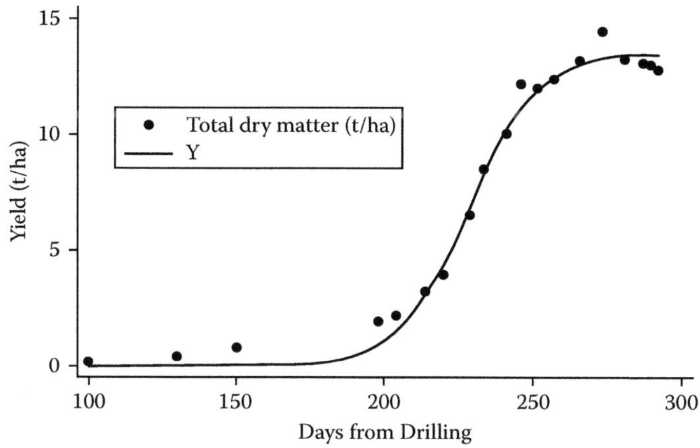


**Figure 10.15** Barley dry matter yield for days after drilling showing the logistics function.

```
3-parameter logistic function, yield = b1/(1 + exp(-b2*(days
- b3)))
----------------------------------------------------------------
    yield |  Coef.    Std. Err.    t  P>|t|    [95% Conf. Interval]
----------+-----------------------------------------------------
      /b1 | 13.54067 .2885517 46.93 0.000    12.92896     14.15237
      /b2 |.0864115 .0082196  10.51 0.000    .0689866     .1038363
      /b3 |227.9791 1.235012 184.60 0.000     225.361     230.5972
----------------------------------------------------------------
```

This logistic function fits well with an $R^2$ value of 0.9964 with all three coefficients having a significant t value (Figure 10.15). The fitted equation is $Y = 13.54067/(1 + e^{-0.0864115(x-227.9791)})$. Enter the following command to see the data and the plotted function:

```
twoway (scatter yield days) (function y = 13.54067/
(1+exp(-.0864115*(x-227.9791))), range (days))
```

It should be pointed out, however, that both commands **nl gom4:** `yield days` and **nl gom3:** `yield days` give reasonably good results as well. This brings up an important point, that fitting data to a specific function should make reasonable sense from a biological perspective. The **nl** command can fit a wide range of functions, but does the chosen function make sense? In this case, any of the three functions would probably be adequate in understanding the underlying data.

# 11

# DATA TRANSFORMATIONS

The analysis of variance has certain underlying assumptions to make the analysis valid. This includes that the data were obtained from a random sample of the population; that the error terms occur randomly, are normally distributed, and are not correlated. The sample populations have equal or homogeneous variances. This is often referred to as *homoscedastic variances*. This can be a little confusing. If this is an analysis of variance, how can the variances be the same? The assumption is that the variance within one group is the same as the variance in other groups. The means of the groups, however, may differ. In ANOVA (analysis of variance), the F-tests are based on a ratio—the variation between the group means divided by the variation within the groups (pooled across groups). It is when there is a disparity between these variances that a significant difference is detected. The variances and treatment means should not be correlated. Finally, the factor levels are assumed to be additive. That is, the model parameters, treatments, replications, error, etc. are added together to create the model. This is often referred to as a linear model or it has linearity.

Not all data will meet these underlying criteria, but often the data can be transformed so that they do. One underlying assumption is the errors are normally distributed. This is the classic bell-shaped curve. There are cases where the data deviate from normality and it may be corrected by transforming the data. Load the dataset Onion Disease Transform Data.dta, which is a dataset from an onion variety trial with data on the number of diseased plants per plot, the number of seedstems (flowering), and the number of doubled bulbs per plot.

Reasonable steps to approaching the problem of analyzing these data would be to run the ANOVA and determine if the residuals (errors) are normally distributed. Enter the following to compute the ANOVA on the raw data:

```
anova plantcount var rep
```

Stata offers a number of commands that can be used to determine normality of the residuals. This includes the **rvpplot** and **rvfplot** commands that were shown in Chapter 10, Correlation and Regression. Data in these scatterplots should appear randomly about 0 on the *y*-axis.

Stata offers additional tests for checking normality including **hettest**, **swilk**, **sfrancia**, **sktest**, and **ladder**. The ladder command also evaluates several transformations to determine if they are normally distributed as well. Enter the following command after the ANOVA has been calculated:

```
hettest
```

which results in the following output:

```
Breusch-Pagan / Cook-Weisberg test for heteroskedasticity
         Ho: Constant variance
         Variables: fitted values of plantcount

         chi2(1)      =   144.96
         Prob > chi2  =   0.0000
```

Using a standard, such as 0.05, for probability; values lower than this would indicate a nonnormal distribution (Prob>chi$^2$). This test indicates that the null hypothesis ($H_o$) should be rejected because the chi$^2$ probability is highly significant. Other commands, such as **swilk** and **sfrancia**, also indicate whether the data are normally distributed. Enter the following commands and see the output:

```
                  swilk plantcount

               Shapiro-Wilk W test for normal data

    Variable |    Obs       W           V          z       Prob>z
-------------+----------------------------------------------------
  plantcount |    120    0.70051     28.819     7.530    0.00000

                 sfrancia plantcount

               Shapiro-Francia W' test for normal data

    Variable |    Obs       W'          V'         z       Prob>z
-------------+----------------------------------------------------
  plantcount |    120    0.69665     32.103     6.938    0.00001
```

The calculated W and W' statistics indicate the data depart significantly from normality with Prob>z of 0.00000 and 0.00001, respectively. These tests also give an indication of the departure from normality with the V and V' values. The median value is 1 for these indices with a normally distributed population. The drawback to these tests is the number of observations must be between 4 ≤ n ≤ 2,000 for the Shapiro–Wilk test and 5 ≤ n ≤ 5,000 for the Shapiro–Francia test. This is really not a drawback in most cases, certainly not in most planned experiments.

Finally, two additional tests to consider are the **sktest** and **ladder**. The former evaluates the skewness and kurtosis for normality combining the two for an overall test of normality. Enter the following command:

```
sktest plantcount seedstems doubles
```

This results in the following output:

```
                Skewness/Kurtosis tests for Normality
                                                   ------- joint ------
Variable    |  Obs  Pr(Skewness) Pr(Kurtosis)  adj chi2 (2)   Prob>chi2
------------+----------------------------------------------------------
plantcount  |  120     0.0000       0.0000        60.57          0.0000
 seedstems  |  120     0.0000       0.0004        38.18          0.0000
   doubles  |  120     0.0000       0.0000          .            0.0000
```

All three variables are significantly different from a normal distribution. The adjusted chi-square is a measure of deviation from normality.

Using the **ladder** command not only calculates the chi-square and probability of the data deviating from normality, but also calculates these values for several transformations. Enter the following command:

```
ladder plantcount
```

This results in the following output:

```
Transformation        formula                    chi2 (2)      P(chi2)
----------------------------------------------------------------------
cubic                 plantc~t^3                        .        0.000
square                plantc~t^2                        .        0.000
identity              plantc~t                      60.57        0.000
square root           sqrt (plantc~t)               26.00        0.000
log                   log(plantc~t)                  1.73        0.420
1/(square root)       1/sqrt (plantc~t)             61.80        0.000
inverse               1/plantc~t                        .        0.000
1/square              1/(plantc~t^2)                    .        0.000
1/cubic               1/(plantc~t^3)                    .        0.000
```

In this table, the identity chi-square is the same as from the previous command and represents the untransformed data. Several transformations are calculated and several are undefined. The log transformation has a chi-square value of 1.73 and a probability of 0.420, which indicates this transformation is normally distributed and would be a good one to use in analyzing the data. The ladder command also can include the option **generate**, which generates a new variable with the transformation that has the smallest chi-square value. It should be noted that not all data can be transformed to normality and, in such cases, other statistical techniques should be considered.

Generate a new variable of plantcount using the natural log transformation

```
generate transpc = log(plantcount)
```

In this particular case, there is not much difference when an ANOVA is done with the original data compared to the transformed data; however, when the CV (coefficient of variance) is examined and the detectable differences between means are examined, there are considerable differences. The CV with the original data was 89% compared to 26% with the transformed data.

Skewness is a measure of an entire distribution that is most notable in the tails of a normal distribution, and kurtosis measures the shape of the normal curve or peakedness. A negative skewness indicates the tail is longer on the left of the distribution with the median lying to the right of the mean, whereas a positive skewness has a longer tail on the right with the median lying to the left of the mean. Kurtosis, which measures peakedness, has a value of 3 for a normal distribution. As this value goes down, the flatter the distribution is, and, conversely, as it goes up, the narrower the distribution. Stata can calculate these values with a couple of different commands. One method is to use the **summarize** command with the **detail** option. Another option is to use the **tabstat** command. Enter the following command:

```
tabstat plantcount transpc, statistics(skewness
kurtosis mean median)
```

This results in the following output:

```
   stats |  plantc~t   transpc
---------+--------------------
skewness |  2.599091 -.1750206
kurtosis |  10.44925  3.341528
    mean |  36.70833  3.157877
     p50 |      25.5  3.238486
------------------------------
```

The column headings include `stats`, `plantc~t` (abbrev. `plantcount`), and `transpc`, which represent the measured statistic, `plantcount`, and transformation of `plantcount`, respectively. Note how the skewness and kurtosis changes with the transformation from `plantcount` to `transpc` representing a more normal distribution. `p50` represents the median, and notice how it is less than the mean with the original data (mean = 36.7) indicating a positive skewness and slightly to the right with the transformed value (`transpc`) (mean = 3.2). Remember the kurtosis will be a 3 with a normal distribution and notice how the transformation is much closer to this value.

The log transformation, which we have used here to reduce skewness and kurtosis for a more normal distribution, also is often used with data where the variances are not homogeneous or they are said to be heteroscedastic and the standard deviations are proportional to the means. To see this with the plantcount, data enter the command

```
anova plantcount var rep
```

This will calculate and display the ANOVA table, and immediately after this is calculated, enter the command

```
rvfplot
```

This will graph the residuals versus the fitted values, which should occur randomly around 0. If the values don't, then this is an indication that the variances are not homogeneous. Calculate the **anova** for the transpc data and then enter the **rvfplot** command.

Figure 11.1 shows this graph for both the `plantcount` and `transpc` data. Notice how the points are clustered at one end of the graph with the untransformed data and the points appear more random after transformation.

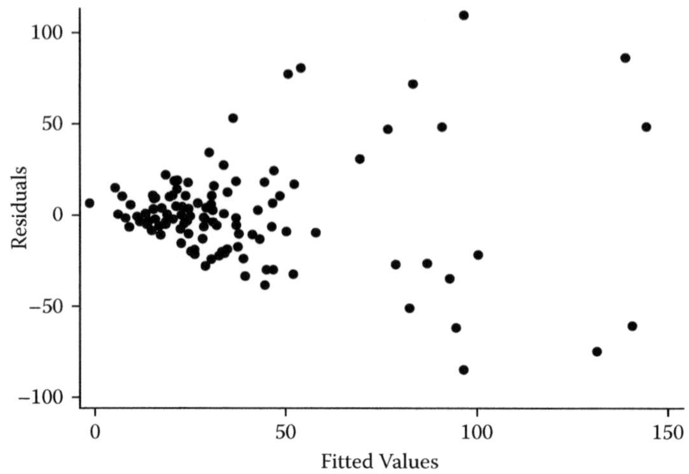


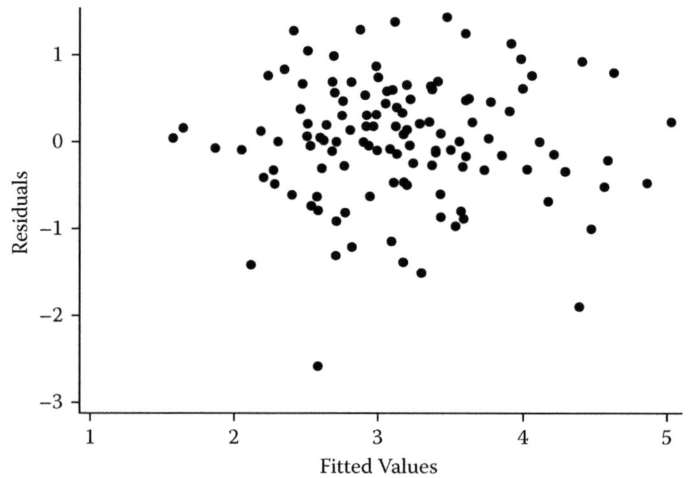


**Figure 11.1** Residual versus fitted values for untransformed data (plantcount) and log transformation (transpc).

Load the dataset Log Transform.dta, which is a dataset from Zar (1974, p. 184) that also illustrates this problem. Create a new column transforming measure to the log of this value by entering the command

```
generate tranmeas = log10(measure + 1)
```

In this case, we are using the log10' function, which is the base 10 log rather than log, as used previously, which is the natural log. Either can be used; however, in this case, to conform to Zar's example, we are using base 10 log. Now enter the command

```
tabstat measure tranmeas, statistics(mean sd var cv)
by(grp)nototal
```

This command calculates the mean, standard deviation, variance, and CV for both the original data (`measure`) and the transformed data (`tranmeas`) as shown below:

```
Summary statistics: mean, sd, variance, cv
  by categories of: grp

     grp |   measure  tranmeas
---------+--------------------
       1 |      3.28  .6306574
         |  .2863564  .0293007
         |      .082  .0008585
         |  .0873038  .0464605
---------+--------------------
       2 |      6.94  .8988198
         |  .6024947  .0329628
         |  .3629999  .0010865
         |  .0868148  .0366734
------------------------------
```

The variances for the original data are obviously different (grp 1 = 0.082 versus grp 2 = 0.3629999) for each treatment and the standard deviations are proportional to the means (grp 1 = 0.2863564 versus grp 2 = 0.6024947) resulting in coefficients of variation that are similar. After transformation, the variances are homogeneous (grp 1 = 0.0008585 versus grp 2 = 0.0010865) and the standard deviations are not proportional to the means (grp 1 = 0.0293007 versus grp 2 = 0.0329628).

Finally a log transformation may be used where the effect is multiplicative rather than additive. For example, in a RCBD (randomized complete block design), it is assumed there is an additive treatment and block effect. That is, from one block to another, the effect does not change in orders of magnitude.

This is another case where we will go outside of Stata to find a command. Enter **findit nonadd** in the Command window while connected to the Internet. This will locate this command, **nonadd**, which can be downloaded and installed in Stata. Load the dataset Onion Disease Transform Data.dta and enter the command

```
nonadd seedstems var rep
```

This calculates Tukey's test for nonadditivity to determine if the assumption of additivity is violated. The results are

```
Tukey's test of nonadditivity (Ho: model is additive)
SS nonadd =    df = 1
F (1,86) = 7.2406072   Pr > F:.00856253
```

In this case, we see that the data differ significantly from being additive. Transform these data with a log transformation and compute Tukey's test again. Enter the commands

```
generate transtem = log10(seedstems+1)
nonadd transtem var rep
```

This results in the following output:

```
Tukey's test of nonadditivity (Ho: model is additive)
SS nonadd =    df = 1
F (1,86) = 1.3522982   Pr > F:.2480934
```

The transformed data now meet the criteria of additivity.

The log transformation can be either a base 10 or natural log or any other log base and the effect will be similar, although a base 10 log will probably work better with data that are multiplicative. Usually some constant is added to the value before this transformation is used particularly if there are any zeros in the dataset or numbers very close to 0. This prevents such data points from being missing data in the transformation. Finally, this type of transformation will not work with negative numbers as these also will be missing data points after the transformation.

Another type of transformation that is commonly used is the arcsine or angular transformation. This type of transformation is often used with percent data particularly when the percentages occur both below 30% and above 70%. These types of datasets often exhibit a binomial distribution rather than a normal distribution and the treatment variances are often less at the extremes of the range than in the middle. This transformation is y = arcsine(square root(x)) where x is the original data.

If you examine statistical textbooks, they usually present such transformations in degrees with the original data often presented as percentages (i.e., 1–100%). The **asin()** function in Stata requires that the input data be in the range of –1 to 1 and the results are presented as radians. This can easily be accommodated by dividing percent data by 100. Load the dataset Lettuce Seed Arcsine Transformation.dta (Little and Hills, 1978, p. 158). This is a dataset of the number of germinating lettuce seeds in samples of 50 seeds. The experimental design was a CRD (completely randomized design).

Examine the dataset by using the **ladder** and **oneway** commands with germ as the dependent variable. The **ladder** command indicates that the dataset is not normally distributed and the **oneway** command calculates an ANOVA with Bartlett's test for equal variance. The assumption that the treatments have equal variance is suspect.

Because this transformation is used with percent data, we will multiple each data point by 2 so the values are on a scale of 0–100. Then divide each by 100, so it is in the range that the **asin()**function requires. Enter the following command:

```
generate trangerm = asin(sqrt(germ*2/100))
```

This generates a new variable with the transformed germination data. Because this is a CRD, we can compare the analysis of the original data and the transformed data with the **oneway** command and see how Bartlett's test differs between them. Enter the commands and see the results.

```
oneway germ trt
```

Original data:

```
                       Analysis of Variance
    Source         SS           df       MS            F      Prob > F
------------------------------------------------------------------------
Between groups  25265.9861   23   1098.52114    148.12       0.0000
Within groups          356   48   7.41666667
------------------------------------------------------------------------
    Total       25621.9861   71   360.873044

Bartlett's test for equal variances:  chi2(23) =  35.6874
Prob>chi2 = 0.044
```

```
oneway trangerm trt
```

Transformed data:

```
                        Analysis of Variance
    Source          SS          df       MS            F      Prob > F
------------------------------------------------------------------------
Between groups  18.1114002   23   .787452182    100.14     0.0000
 Within groups  .377453807   48   .007863621
------------------------------------------------------------------------
    Total        18.488854   71   .260406394

Bartlett's test for equal variances:  chi2(23) =   9.6640
Prob>chi2 = 0.993
```

Compare the two ANOVA tables and notice how the $chi^2$ is no longer significant ($p \leq 0.05$) with the transformed data indicating the variances are equal. In both analyses the treatment (between groups) effects are significant; however, the detected differences between the treatments will be different. In Chapter 8, Post Hoc Tests, I covered multiple range tests including Duncan's Multiple Range Test, which we will use again here. The **pwcompare** command will give us similar results, but all comparisons are shown, whereas the duncan.do file condenses the output making it easier to see the results. Load the do file duncan.do. This program was originally written to analyze data from a RCBD, so a couple of minor changes will be needed to use it with a CRD. Find the following piece of code and make the following changes; comment out the rep argument in the first line below as well as e(df_2)+1 in the third line. Enter 3 as the value for the local macro repl.

```
anova `depend' `indep' `rep'        // Calculates anova
local var = (e(rmse))^2             // Error mean square from ANOVA
local repl = e(df_2)+1              // Number of replications
```

When finished, it should look like this:

```
anova `depend' `indep' //`rep'      // Calculates anova
local var = (e(rmse))^2             // Error mean square from ANOVA
local repl = 3 //e(df_2)+1          // Number of replications
```

Run the duncan.do file and then enter the command

```
duncan germ trt
```

followed by

```
duncan trangerm trt
```

Below is part of the output including the means from the first analysis and the letters used to separate the means from both the first and second analysis.

| | | |
|---|---|---|
| 49.33 | a | a |
| 49 | ab | a |
| 48.33 | ab | ab |
| 47.67 | ab | ab |
| 45.33 | abc | bc |
| 45 | abc | bc |
| 43.67 | bc | bc |
| 42 | c | c |
| 41.67 | c | c |
| 41 | c | c |
| 30.67 | d | d |
| 27.33 | de | d |
| 24.33 | e | de |
| 18.33 | f | ef |
| 17 | fg | ef |
| 16.33 | fg | ef |
| 12.67 | gh | fg |
| 8 | h | g |
| 1.67 | i | h |
| 0.67 | i | h |
| 0.67 | i | h |
| 0.33 | i | h |
| 0.33 | i | h |
| 0.33 | i | h |

Notice the difference in Duncan's Multiple Range Test between the two analyses.

Normally, the transformed means will be back transformed to the original units after the analysis and these back transformed means can often be slightly different from the means calculated on the original data. The back transformation is done in reverse order of the transformation, thus, the outermost calculation is done first working inward through the parenthesis to the innermost calculation. The back transformation of the seed data is

(sin(trangerm)^2)/2*100

**Table 11.1** Seed treatment means from the original data and the back transformed means

| TREATMENTS | ORIGINAL DATA MEANS | BACK TRANSFORMED MEANS |
|---|---|---|
| 1 | 0.3333333 | 0.1117765 |
| 2 | 0.3333333 | 0.1117765 |
| 3 | 0.3333333 | 0.1117765 |
| 4 | 0.6666667 | 0.224912 |
| 5 | 0.6666667 | 0.224912 |
| 6 | 1.666667 | 1.110808 |
| 7 | 8.0 | 7.95229 |
| 8 | 12.66667 | 12.63044 |
| 9 | 16.33333 | 16.28491 |
| 10 | 17.0 | 16.90513 |
| 11 | 18.33333 | 18.21566 |
| 12 | 24.33333 | 24.3327 |
| 13 | 27.33333 | 27.34132 |
| 14 | 30.66667 | 30.7889 |
| 15 | 41.0 | 41.01451 |
| 16 | 41.66667 | 41.83367 |
| 17 | 42.0 | 42.13401 |
| 18 | 43.66667 | 43.91429 |
| 19 | 45.0 | 45.30207 |
| 20 | 45.33333 | 45.63686 |
| 21 | 47.66667 | 47.82714 |
| 22 | 48.33333 | 48.37023 |
| 23 | 49.0 | 49.34827 |
| 24 | 49.33333 | 49.55389 |

In Table 11.1 are the treatment means from the original data and the treatment means from the back transformed data.

Other common transformations include using the reciprocal of the data or squaring the data. Any transformation can be used, as long as it is applied to all the data points. It should be emphasized, however, that transformations are used to correct violations of the underlying assumptions in the analysis, not as a fishing expedition to find the results you want.

# 12

# Binary, Ordinal, and Categorical Data Analysis

In previous chapters, we have dealt primarily with continuous data, such as yield. There are occasions where data are not continuous. For example, some data will have only two possible values, such as whether a plant is diseased or healthy. Other examples are sex, alive or dead, etc. It may be useful in such cases to know the probability of a specific ratio of events. For example, sex ratios between males and females is approximately 50%. Not every sample from a population is going to have exactly half male and half female individuals, however. If you took a sample of 20 individuals, it would not be unusual to have 9 males and 11 females and, although a rarer event, it is even possible to have all 20 of the individuals be either male or female. Such binomial events can be calculated. Open the Binomial.dta dataset and enter the command

```
bitest sex ==.5, detail
```

This results in the following output:

```
 Variable | N    Observed k    Expected k    Assumed p    Observed p
----------+-----------------------------------------------------------
      sex | 20        4             10         0.50000       0.20000

  Pr(k >= 4)             = 0.998712  (one-sided test)
  Pr(k <= 4)             = 0.005909  (one-sided test)
  Pr(k <= 4 or k >= 16)  = 0.011818  (two-sided test)

  Pr(k == 4)             = 0.004621  (observed)
  Pr(k == 15)            = 0.014786
  Pr(k == 16)            = 0.004621  (opposite extreme)
```

Datasets of this type can only have data as either 0 or 1, representing binomial data. In this case, it could be interpreted that 1 is female and 0 is male. Look at the value Pr(K == 4), which is 0.004621. This is the probability of selecting a sample of 20 individuals and having only 4 be females. This output also indicates what the probability is for selecting a sample with 4 or less females (p = 0.005909) as well as selecting a sample with 4 or more females (p = 0.998712).

Although calculating probabilities of such binomial data is interesting, it is probably of greater value to the researcher to see if collected data meet some underlying binomial ratio. In fact, this idea of evaluating data for a specific ratio can be expanded to more than just two categories.

Zar (1974, p. 285) has an example of such a problem. Data on 54 litters of 5 offspring are presented with the question: Do these litters meet the criteria of 50% female and 50% male offspring? Load the dataset Offspring.dta. Using the **bitesti** command, the probability of having litters with 0, 1, 2, 3, 4, or 5 female offspring can be calculated. The **bitesti** command is the immediate form of the **bitest** command. In other words, it does not require a dataset to be used and can calculate such probabilities directly. Enter the command

```
bitesti 5 0 0.5, detail
```

This command calculates what the probability is of having 0 females in a litter of 5 individuals when the underlying ratio is 50% females (0.5). The **detail** qualifier calculates the probability of having 0 females in a litter of 5 offspring, which is 0.031250. See the `Pr(k == 0) = 0.031250 (observed)` line in the detailed output. This command can be entered for each number of female offspring (i.e., 0, 1, 2, 3, 4, 5) in a litter of 5, but that can get tedious. A different approach would be to set up a do-file to calculate these values, which are used to determine if the litter samples fit the 1:1 ratio of males to females. The **chi2tail()** density function is used to calculate the appropriate value, but a user written program is available that may be more useful in this case. Enter the following command while connected to the Internet:

```
findit csgof
```

This command will search the Internet for the specified command (i.e., **csgof**), which calculates a chi² ($\chi^2$) goodness-of-fit. Once you have found **csgof**, download it so that it is available to use in Stata. It also includes a help file that explains how to use the command.

The chi² ($\chi^2$) goodness-of-fit is calculated by the following formula:

$$\chi^2 = \sum_{i=1}^{k} \frac{(OB - Ex)^2}{Ex}$$

This formula is the sum of the observed (Ob) minus the expected (Ex) squared over the expected. This value, along with the degrees of freedom, is used to calculate a probability.

Open the do-file Binomial `distribution.do`, which calculates the probabilities for each possible ratio among the litters. Enter the command

```
chisquare offspring
```

This information is then used with the **csgof** command to calculate the chi² and the probability with the following output:

```
+-------------------------------------+
| offspr~g   expperc   expfreq  obsfreq |
|-------------------------------------|
|        0     3.125    1.6875        3 |
|        1    15.625    8.4375       10 |
|        2     31.25    16.875       14 |
|        3     31.25    16.875       17 |
|        4    15.625    8.4375        9 |
|-------------------------------------|
|        5     3.125    1.6875        1 |
+-------------------------------------+

chisq(5) is 2.12, p = .8325
```

These results, with a p value of 0.8325, indicate that the makeup of the litters does not deviate appreciably from the expected ratio of 50% females and 50% males.

Finally, the Binomial distribution.do file is specifically written for the Offspring.dta dataset. It could be modified, however, to have a more general application. This may be a good exercise to improve your programming skills.

The chi² goodness-of-fit calculation has other applications most notably in genetics. Many characteristics are inherited by just one or two genes. For example, in simple Mendelian inheritance, with a plant species, where a gene has a single dominant allele and a homozygous individual with this gene is crossed with a homozygous recessive individual, all of the individuals in the first generation ($F_1$) will exhibit the dominant characteristic. However, when these crossed individuals are selfed (crossed with themselves), the second generation ($F_2$) will segregate into a 3:1 ratio with ¾ of the individuals exhibiting the dominant gene and ¼ exhibiting the recessive gene.

Load the dataset Watermelon ZYMV Resistance.dta. This is a dataset of two variables. The first variable (f2zymv) is the scoring of the $F_2$ generation for resistance to Zucchini Yellow Mosaic Virus. The second variable is the scoring of the $F_1$ generation backcrossed to the recessive parent. Enter the command

```
csgof f2zymv, expperc(25 75)
```

The expperc (expected percentages) must add to 100% and the order they are entered is important because this command sorts the values in descending order before doing the calculations. Reversing the order of 25 and 75 will result in erroneous results. The results are

```
+-------------------------------------+
| f2zymv   expperc   expfreq   obsfreq |
|-------------------------------------|
|      0        25        54        50 |
|      1        75       162       166 |
+-------------------------------------+

chisq(1) is .4, p = .5297
```

The results indicate that the data do indeed meet the expected 3:1 ratio. Enter the **csgof** command for the backcross data with expected ratio of 1:1. To do this, enter the following and see the results:

```
csgof bc1zymv, expperc(50 50)
```

```
+-----------------------------------------+
| bc1zymv   expperc   expfreq   obsfreq |
|-------------------------------------|
|       0        50        56        51 |
|       1        50        56        61 |
+-------------------------------------+

chisq(1) is .89, p = .3447
```

The **expperc(50 50)** is equivalent to a 1:1 ratio and the observed frequencies do not differ from this ratio, p = 0.3447.

Finally, when dealing with ratios of just two categories, as with these simple inheritances where the degrees of freedom will be 1, using Yates' correction will give more accurate results. Yates' correction is nothing more than subtracting 0.5 as in the following equation:

$$\chi^2 = \sum_{i=1}^{k} \frac{\left(\left|OB - Ex\right| - 0.5\right)^2}{Ex}$$

To incorporate this modification to the **csgof** command enter

```
which csgof
```

This will give you the pathname to this command, which you should copy. Then select the View... menu item under the File menu and paste the pathname in the dialog box. This will open a Viewer window with a list of the **csgof** ado' command. Select this entire file and copy it into a Do-File Editor window. At this point you can make modifications to this file. You can run these modifications and use the modified command immediately or you can save the file for later use. I would suggest that you not overwrite the original **csgof** command, but store the modified file elsewhere. It is standard and safe advice to rename the modified file/command. This also will require changing the **program** lines in the file. To change the **program** lines find the following and change them as indicated:

Original lines:

```
capture program drop csgof
program define csgof
```

New lines:

```
capture program drop csgofy
program define csgofy
```

Find the following line in the csgof file:

```
quietly gen `chi1' = (obsfreq - expfreq)^2/expfreq
```

and make the following modification:

```
quietly gen `chi1' = (abs(obsfreq - expfreq)-0.5)^2/expfreq
```

This then has incorporated Yates' correction. In this particular case, it has not changed the outcome, but there will be cases where the outcome will be different. The original program (csgof) stands for chi-square goodness of fit and in the modified file adding y is to indicate Yates' correction has been added. At this point, you would want to save the file under the new name, csgofy.ado.

As you might expect, there can be more than two categories in a goodness-of-fit test. Davis (2000, p. 150) has an example for tractor repairs. In his example, there are five different makes of tractor for which data were collected on the types of repairs. The repairs were cataloged as either electrical, fuel supply, or other. Open the dataset Tractor Repair.dta, which encompasses these data. There are two variables: tractor, which are the five makes of tractors labeled 1–5, and repair, which indicates the repairs for each tractor (1: electrical, 2: fuel supply, or 3: other). Enter the following command:

```
tabulate tractor repair, chi2 expected
```

This results in the following output:

```
+--------------------+
| Key                |
|--------------------|
|     frequency      |
| expected frequency |
+--------------------+

      Five |
 different |   Repairs: 1=Electrical, 2=Fuel supply,
  makes of |              3=Other
   tractor |         1           2           3 |     Total
-----------+-----------------------------------+----------
         1 |        17          19           7 |        43
           |      14.4        18.7         9.9 |      43.0
-----------+-----------------------------------+----------
         2 |        14           7           9 |        30
           |      10.0        13.0         6.9 |      30.0
-----------+-----------------------------------+----------
         3 |         6          21          12 |        39
           |      13.1        17.0         9.0 |      39.0
-----------+-----------------------------------+----------
```

```
         4 |        33          44          19 |         96 
           |      32.1        41.7        22.1 |       96.0 
-----------+-----------------------------------+----------
         5 |         7           9           6 |         22 
           |       7.4         9.6         5.1 |       22.0 
-----------+-----------------------------------+----------
     Total |        77         100          53 |        230 
           |      77.0       100.0        53.0 |      230.0 

          Pearson chi2 (8) =  12.9152   Pr = 0.115
```

This table lists the five tractors in the first column and then lists the frequency of repairs for each (e.g., tractor 1: 17 electrical, 19 fuel supply, and 7 other). The second set of numbers (14.4, 18.7, 9.9) is the expected frequencies for each category. The expected frequencies are calculated by

$$\text{Expected frequency} = \frac{\text{row total} \times \text{column total}}{\text{grand total}}$$

For example, (43)(77)/(230) = 14.4. The `Pearson chi2 (8)` is the $chi^2$ calculation with 8 degrees of freedom (r-1)(c-1) = (5-1)(3-1). The calculated value (12.9152) and the probability (Pr = 0.115) indicate that the make of tractor and number of repairs are independent. To put it another way, there isn't any difference in repair frequency due to make of tractor.

Although this datum is entered individually for each tractor, it might have been compiled with the repair frequency listed for each tractor type. To see how this might have been entered, enter the following commands:

```
preserve
contract repair tractor, freq(number)
```

This contracts the dataset using both the repair and tractor variables to compile a third variable, frequency, with the frequency of each tractor/repair combination. The option freq(frequency) can be left off and Stata will automatically create the new variable with _freq as the new variable name. This shows you another method of entering such data with frequencies rather than each tractor individually. Next, enter the following commands:

```
tabulate tractor repair [fweight = number], chi2 expected
restore
```

The **tabulate** command entered with (**fweight** = number) uses the number variable as a frequency weight to calculate the output, which is exactly the same as shown above. The **preserve** and **restore** commands preserve and restore the original dataset after using the **contract** command.

In previous chapters, the data generally had to meet certain underlying criteria, such as normality, additivity, homogeneous variances, etc. In some cases, it was possible to transform data to meet these criteria. There are also tests that do not require these underlying assumptions. These methods are often referred to as nonparametric tests. In general, if the underlying assumptions for the parametric tests are true, then these nonparametric tests are not as powerful.

One such test is called the Sign Test. The Sign Test is much like the paired t-test without any underlying assumptions about the populations. This test is an evaluation of medians to see if two medians differ significantly from 0. In general, this test works better with 20 or more paired data points. The advantage of not having specific requirements in the population is offset by the loss of information concerning the magnitude of the differences. Open the dataset Sign Test Food Products.dta. This is a dataset of 22 people rating two different snacks on a scale of 1–20 with 20 considered the best (Davis, 2000, p. 198). There is an error in the dataset as listed in Davis' text, so the results will be different here. Enter the following command:

```
signtest tomato = apricot
```

This results in the following output:

```
Sign test

        sign |    observed    expected
-------------+------------------------
    positive |           7          11
    negative |          15          11
        zero |           0           0
-------------+------------------------
         all |          22          22
```

```
One-sided tests:
  Ho: median of tomato - apricot = 0 vs.
  Ha: median of tomato - apricot > 0
      Pr(#positive >= 7) =
         Binomial(n = 22, x >= 7, p = 0.5) =  0.9738

  Ho: median of tomato - apricot = 0 vs.
  Ha: median of tomato - apricot < 0
      Pr(#negative >= 15) =
         Binomial(n = 22, x >= 15, p = 0.5) =  0.0669

Two-sided test:
  Ho: median of tomato - apricot = 0 vs.
  Ha: median of tomato - apricot != 0
      Pr(#positive >= 15 or #negative >= 15) =
         min(1, 2*Binomial(n = 22, x >= 15, p = 0.5)) =  0.1338
```

The results are presented with probabilities for equal medians, with one median greater than the other and, finally, with one median less than the other. Which of these results to use is dependent on the data and what specifically the experiment is about. In this particular case, the two-sided test is the appropriate analysis because we are not interested in one particular snack being less than or greater than the other. In this case, with a probability of 0.1338, the medians do not differ from one another or, to put it another way, the difference between the medians do not differ from 0.

There are cases where the one-sided test is going to be more appropriate. For example, load the dataset Heifer Vitamin A.dta. This is a dataset of heifers paired for size to examine the effect of vitamin A on weight gain (Steel and Torrie, 1980. p. 98). Enter the following command:

```
signtest control = vitamina
```

with the following results:

```
Sign test

        sign |    observed    expected
-------------+------------------------
    positive |           4           7
    negative |          10           7
        zero |           0           0
-------------+------------------------
         all |          14          14
```

```
One-sided tests:
  Ho: median of control - vitamina = 0 vs.
  Ha: median of control - vitamina > 0
      Pr(#positive >= 4) =
         Binomial(n = 14, x >= 4, p = 0.5) =  0.9713

  Ho: median of control - vitamina = 0 vs.
  Ha: median of control - vitamina < 0
      Pr(#negative >= 10) =
         Binomial(n = 14, x >= 10, p = 0.5) =  0.0898

Two-sided test:
  Ho: median of control - vitamina = 0 vs.
  Ha: median of control - vitamina != 0
      Pr(#positive >= 10 or #negative >= 10) =
         min(1, 2*Binomial(n = 14, x >= 10, p = 0.5)) =  0.1796
```

In this case, the second one-sided test is appropriate because we are interested in whether vitamin A increases weight gain. In this case, with a probability of 0.0898, the gain in weight was not significant at the 5% level. Interestingly, the paired t-test does indicate a significant difference showing the greater power in the t-test. Finally, it is possible to use the **signtest** with a specific value, such as **signtest** control=250.

Another nonparametric test that can be used is Wilcoxon's Signed Rank Test. This test also may be referred to as Wilcoxon's Paired Sample Test or Wilcoxon's Matched Sample Test. This test is considered somewhat better than the Sign Test because it takes into account the magnitude of the differences. The only requirement for the data in this case is that it be symmetrical. In rare cases, the data may have to be transformed if highly skewed. This test is used with data similar to what is used with a paired sample t-test. The Wilcoxon's Signed Rank Test will not be as powerful as the paired t-test when the data reasonably match the assumptions for the t-test, but it is often used with ordinal scale data. This test can be used with paired data or against a single data point. For an example of the latter application, open the dataset Linseed Variety Yields.dta, which is a dataset of yields from a new linseed variety grown at various locations in southeast England (Clewer and Scarisbrick, 2001, p. 296). The standard linseed variety for this area produces 2 t/ha. Enter the following command and see the results:

```
signrank yield = 2
```

```
Wilcoxon signed-rank test

        sign |      obs   sum ranks    expected
-------------+---------------------------------
    positive |        7        47.5          27
    negative |        2         6.5          27
        zero |        1           1           1
-------------+---------------------------------
         all |       10          55          55

unadjusted variance         96.25
adjustment for ties         -0.25
adjustment for zeros        -0.25
                       ----------
adjusted variance           95.75

Ho: yield = 2
             z =   2.095
    Prob > |z| =   0.0362
```

The median for this dataset is 2.35 t/ha with a probability > |z| of 0.0362, which indicates the median is significantly different from the 2.0 t/ha of the standard linseed variety. This example also shows that this test (**signrank**) is more powerful than the Sign Test (**signtest**), which has a probability of 0.1797 for the same dataset.

Like the Sign Test, the Wilcoxon's Signed Rank Test also can be used with paired data. With the dataset Fungi Paired Test.dta, enter the following command and see the results:

```
signrank fungusa = fungusb

Wilcoxon signed-rank test

        sign |      obs   sum ranks    expected
-------------+---------------------------------
    positive |        3        11.5          27
    negative |        6        42.5          27
        zero |        1           1           1
-------------+---------------------------------
         all |       10          55          55

unadjusted variance         96.25
adjustment for ties         -0.25
adjustment for zeros        -0.25
                       ----------
adjusted variance           95.75

Ho: fungusa = fungusb
             z =  -1.584
    Prob > |z| =   0.1132
```

This is a dataset of 10 tomato plants where fungal strains A and B are inoculated on two different leaves on each plant (Clewer and Scarisbrick, 2001, p. 298). The number of fungal colonies that develop is then counted. In this case, with a probability > |z| equal to 0.1132, the null hypothesis that there are no differences between the fungal strains cannot be ruled out.

The Mann–Whitney Test is a nonparametric procedure for unmatched data. In this case, it is the ranking of the data that is important to the analysis rather than the actual data. Although this is a nonparametric approach, there are some underlying assumptions that the populations are similar, but not necessarily normal with similar variances. This test is the nonparametric equivalent of the independent sample t-test. An experiment was conducted comparing a standard wheat variety to a new variety in a CRD (completely randomized design) with 10 plots for the standard variety and 6 plots for the new variety (Clewer and Scarisbrick, 2001, p. 300). Open the dataset Wheat Variety Test.dta and enter the following command to see the results:

```
ranksum yield, by(variety)

Two-sample Wilcoxon rank-sum (Mann-Whitney) test

     variety |      obs    rank sum    expected
-------------+---------------------------------
           1 |        6          71          51
           2 |       10          65          85
-------------+---------------------------------
    combined |       16         136         136

unadjusted variance        85.00
adjustment for ties        -0.88
                       ----------
adjusted variance          84.12

Ho: yield(variety==1) = yield(variety==2)
             z =  -2.181
    Prob > |z| =   0.0292
```

The results indicate that the null hypothesis of equal yield should be rejected with a Prob > |z| = 0.0292. The median of the new wheat variety is 2.35 t/ha, while the standard variety had a median of 2.0 t/ha.

The Kruskal–Wallis test is a nonparametric test with independent samples where more than two medians are involved. It is similar to the one-way ANOVA (analysis of variance) and has been referred to as an analysis of variance with ranks.

The dataset Plant Flies.dta is a dataset of the number of flies per square meter of foliage collected from a forest at different heights (herbs, shrubs, and trees) (Zar, 1974, p. 140). Open this dataset and enter the following:

```
kwallis flies, by(plant)
```

```
Kruskal-Wallis equality-of-populations rank test

  +------------------------+
  | plant  | Obs  | Rank Sum |
  |-------+-----+----------|
  |      1 |    4 |    41.00 |
  |      2 |    4 |    23.00 |
  |      3 |    4 |    14.00 |
  +------------------------+

chi-squared =      7.269 with 2 d.f.
probability =      0.0264

chi-squared with ties =      7.269 with 2 d.f.
probability =      0.0264
```

The probability of 0.0264 indicates that there are differences in the number of flies between the different strata. The medians are 10.85, 6.95, and 5.55 for the herbs, shrubs, and trees, respectively.

Let's examine another example using the Kruskal–Wallis test. Open the dataset Rice Insecticides.dta, which is a CRD examining different insecticide treatments to control brown planthoppers and stem borers in rice (Gomez and Gomez, 1984, p. 14). This experiment would normally be analyzed with a one-way ANOVA. In this case, we are going to change a couple of the entries so that it includes some ties. Change the Azodrin treatment with 2387 kg/ha to 2385 kg/ha and change the Dol-Mix (1 kg) treatment with 2537 kg/ha to 2536 kg/ha. This gives us two values that are tied in the dataset. Enter the following command and see the results:

```
kwallis yield, by (trt)

Kruskal-Wallis equality-of-populations rank test

  +---------------------------------+
  |            trt | Obs | Rank Sum |
  |----------------+-----+----------|
  | Dol-Mix (1 kg) |   4 |    65.50 |
  | Dol-Mix (2 kg) |   4 |    97.00 |
  |    DDT + y-BHC |   4 |    92.00 |
  |        Azodrin |   4 |    63.50 |
  |  Dimecron-Boom |   4 |    41.00 |
  |----------------+-----+----------|
  |  Dimecron-Knap |   4 |    35.00 |
  |        Control |   4 |    12.00 |
  +---------------------------------+

chi-squared =     21.050 with 6 d.f.
probability =      0.0018

chi-squared with ties =     21.061 with 6 d.f.
probability =      0.0018
```

Look at these results compared to the previous Kruskal–Wallis analysis. The chi-square with ties in the second analysis is slightly different than the chi-square without ties (21.050 vs. 21.061). In the first analysis, these values are the same (7.269). Because we are dealing with ranked data, ties will have an impact on the outcome.

Friedman's Test is a nonparametric test for two-way classified data, such as would be found in an RCBD (randomized complete block design) where the treatments are one factor and blocking or replications are another. This test is not directly available in Stata, but can be downloaded as an ado file. Use the **findit** command to locate the **friedman** command (remember, you have to have an Internet connection). Install this program and open the dataset Flaxseed Oil Content 2.dta. This dataset is of an experiment evaluating Redwing flaxseed oil content based on growth stage of inoculation with *Septoria linocola,* the causal organism of pasmo (Steel and Torrie, 1980, p. 547).

The **friedman** command, unlike most commands in Stata, requires that each block or replication occur as a separate variable. This is often the way such data are presented in textbooks, rather than the convention of blocking or replication occurring as a single variable label. After opening the dataset, enter the command and see the results.

```
friedman block*

Friedman = 11.0714
Kendall  = 0.5536
p-value  = 0.0500
```

Notice how the variable (`block*`) is entered. Because each variable has the same root name (`block`), adding the asterisk indicates to the command to use all variables with this root. The command could have been entered as **`friedman`** `block1 block2 block3 block4`, and would have resulted in the same outcome. The $chi^2$ value is labeled as Friedman, which is 11.0714. The p-value of 0.0500 indicates there are differences between treatment medians at the 5% level. The ANOVA for these data was highly significant with an F value of 4.83 and a p-value of 0.0080, again showing the greater power of parametric tests when the data do not grossly violate the assumptions. The Kendall number is a value between 0 and 1. The Kendall W, or coefficient of concordance, is an indicator of how well the blocks agree in their ranking of the treatments with higher values indicating greater agreement.

As mentioned throughout this chapter, nonparametric statistics are generally not as powerful as parametric statistics. They are useful, however, where the underlying samples don't meet the requirements of parametric tests and a reasonable transformation is not available.

# Appendix

This is an explanation of a manual method to calculate the adjusted treatment mean square and effective error mean square with the balanced lattice design of Chapter 5. These calculations have been put together in a do-file called ballatadj.do.

The saved scalars from the ANOVA (analysis of variance) estimation command can be used to calculate the necessary adjustment term μ. This is calculated as

$$\mu = \frac{\text{Block}(\text{adj.})\,\text{MS} - \text{Intrablock error MS}}{k^2\left[\text{Block}(\text{adj.})\,\text{MS}\right]}$$

Substituting in this equation results in the following calculation:

$$0.0358982 = \frac{758.789167 - 322.9625}{16\left[758.789167\right]}$$

The following equation entered in the Stata command window will calculate this adjustment:

```
localu = ((e(ss_3)/e(df_3)) - (e(rmse)^2))/
((e(df_1)^2) * (e(ss_3)/e(df_3)))
```

This looks more complicated than it is. The scalars saved from the **anova** estimation command do not include the mean square values, but do include the sum of squares and degrees of freedom, which can be used to calculate the mean square values. To see all the scalars from the ANOVA estimation command, type **ereturn list** immediately after invoking the **anova** command. To calculate the block, adjusted mean square, the `e(ss _ 3)`, which is the block adjusted sum of squares (11,381.8375), is divided by `e(df _ 3)`, the block adjusted degrees of freedom (15). The `e(rmse)` scalar is the root mean square error, which is the square root of the residual mean square.

The **local** u in the equation saves the results of the calculation in a local macro. To see the value of this calculation, enter **display `u'**, which should be 0.0358982. Remember the open and closed single quotes are required to display the value of u.

At this point, to continue the analysis requires creating a new dataset. To begin with, we need to calculate the treatment totals. Start by entering the command **preserve**. This will save the current dataset before generating a new dataset. Enter the command

```
collapse (sum) tiller, by(trt)
```

This will add the tiller values by each treatment creating a new dataset. After entering this command, you can open the Data Editor to see the result. At this point, save this dataset as trttotals.dta or use some other easily remembered name. Now enter **restore**, which will restore the original dataset. Again, enter the **preserve** command to save the current dataset. Reenter the **collapse** command as

```
collapse (sum) tiller, by(block)
```

This results in summing the tiller values by the block variable. This dataset of block totals has to be expanded to the original dataset size of 80 observations and sorted to match the original dataset. To do this, enter the following commands:

```
generate id = 1
expandcl 4, cluster(id generate (ident)
sort ident block
```

The first command generates a new variable of observations all with the value 1. The **expandcl** command expands the dataset to the number of clusters (4), which is indicated by **cluster**(id), in this case a single value. If the id variable had more than one value, this command would assume that each value of id represented a different cluster. The dataset has 20 observations all with the same value for id, hence, the 4 expands the dataset to 80 observations. The **generate**(ident) option, which is part of the **expandcl** command, generates a new variable with each new observation labeled 1–4. Finally, when the dataset is sorted by the ident and block variables, the dataset consists of 80 observations sorted into four groups of the block totals.

This dataset is now ready to merge with the treatments from the original dataset. Enter the command

```
merge 1:1 _n using "Lattice design.dta", keepusing(trt)
```

This merges the trt variable from the Lattice design.dta into the current dataset in memory (to see this dataset, it is available on the disk as Block Treatment merge.dta). The **merge 1:1 _ n** command indicates it is a 1:1 merge by observations. If you open the Data Editor, you will notice another variable called _ merge. This variable indicates if the corresponding observation is from the dataset in memory (called the master dataset) or from the dataset on disk (called the using dataset). If the value is 1, then it is from the master dataset, and if it is 2, it is from the using dataset. The value in this case should all be 3 indicating the observations are from both the master and using datasets. There also can be 4 or 5 indicating missing updated or nonmissing conflict, respectively. Now the dataset should be collapsed again as follows:

```
collapse (sum) tiller, by(trt)
```

The name of the variable tiller should be changed to something else so that it does not conflict with the next command, e.g., btiller. This is important so the next command will work properly. To change the name, enter

```
rename tiller btiller
```

Again, we will merge this dataset with one we created earlier. Enter the command

```
merge 1:1 _n using trttotals.dta
```

Notice a couple of differences in how the **merge** command is used here compared to previously. There are no quotes around the file name because they are only needed if a file name contains spaces. In addition, there is no option **keepusing()** because the entire file on the disk is being merged into the dataset in memory, not just a particular variable. Opening the Data Editor will show a dataset of the 16 treatments with the treatment totals as well as the block totals for each treatment. The _ `merge` variable can be dropped with

```
drop _merge
```

This dataset consists of the totals for each treatment and the block totals for each treatment. For example, the total for treatment 2 adds the values (see Chapter 5, Table 5.2)

152 + 155 + 130 + 152 + 205 = 794

For the block totals, each block in which treatment 2 occurs is added. Treatment 2 occurs in blocks 1, 6, 10, 14, and 18. For block 1, add all the experimental units in this block (see Chapter 5, Table 5.2):

147 + 152 + 167 + 150 = 616

For all the blocks in which treatment 2 occurs, add the block totals:

616 + 586 + 654 + 724 + 742 = 3,322

At this point, a series of commands are entered to generate several new variables and local marcos, which will be used to calculate the adjusted treatment mean square and the adjusted error mean square. The adjustment factor `u`, which was previously calculated, is used in these commands. Because `u` is calculated from the scalars of the most recent **anova** command, you may wish to check if it is still valid. To do this enter

```
display `u'
```

If the adjustment value is not displayed, you will need to reenter the **anova** command and recalculate this value. This value is 0.0358982 and can be substituted in the subsequent calculations if you do not want to redo the **anova**.

The ratio of these adjusted values will be used to calculate the F value and probability. Enter the following commands:

```
count
local k = sqrt (r(N))
gen y = sum (tiller)
local G = y[_N]
gen W = `k' * tiller - (`k'+1) * btiller + `G'
gen T = tiller + `u' * W
gen M = T/(`k' + 1)
gen T2 = T^2
gen y2 = sum (T2)
local TT2 = y2[_N]
local adjTMS = 1/((`k'+ 1) * (`k'^2 - 1)) * (`TT2'
- (`G'^2/`k'^2))
local EEMS = (e(rss)/e(df_r)) * (1 + `k' * `u')
local adjF = `adjTMS'/`EEMS'
display `adjF'
display Ftail(`k'^2-1, (`k'-1) * (`k'^2 - 1), `adjF')
```

The count command counts the number of observations in the current dataset, which is 16. More importantly, it stores this value in the scalar r(N). The next command creates a local macro, k, which is the square root of the scalar r(N) resulting in 4. The next command generates a new variable y that is a running total of the tiller variable. G is a local macro of the last value in the y variable, which has a value of 13,746. The next variable generated is W, which is calculated using the k and G macros with the tiller and btiller variables. The next variables, T, M, T2, and y2, are generated in a like fashion. The TT2 macro is before the last item in the y2 variable. The adjTMS holds the adjusted treatment mean square and the EEMS is the calculated adjustment to the error mean square (effective error mean square). The F value is calculated as the adjusted treatment mean square divided by the EEMS. Finally, the calculated F and the related probability are displayed.

The **`Ftail()`** function calculated the upper tail cumulative F distribution based on the numerator and denominator degrees of freedom and the calculated F value. In this case, the numerator is 15 and the denominator is 45, while the F value is 4.3323959. The probability associated with this F value is 0.00006534, which is highly significant.

# References


Clewer, A. G., and D. H. Scarisbrick. 2001. *Practical statistics and experimental design for plant and crop science*. New York: John Wiley & Sons.

Davis, B. 2000. *Introduction to agricultural statistics*. Albany, NY: Delmar Thomson Learning.

Gomez, K. A., and A. A. Gomez. 1984. *Statistical procedures for agricultural research*, 2nd ed. New York: John Wiley & Sons.

Little, T. M., and F. J. Hills. 1978. *Agricultural experimentation: design and analysis*. New York: John Wiley & Sons.

Palaniswamy, U. R., and K. M. Palaniswamy. 2006. *Handbook of statistics for teaching and research in plant and crop science*. Binghamton, NY: The Haworth Press, Inc.

Steel, R. G. D., and J. H. Torrie. 1980. *Principles and procedures of statistics*. New York: McGraw-Hill.

Zar, J. H. 1974. *Biostatistical analysis*. Englewood Cliffs, NJ: Prentice Hall.

# Index

B

S

## W



## X



## Y



## Z

For Product Safety Concerns and Information please contact our EU representative GPSR@taylorandfrancis.com Taylor & Francis Verlag GmbH, Kaufingerstraße 24, 80331 München, Germany

Batch number: 10725802

Printed by Printforce, the Netherlands